MANAGEMENT OF INDOOR AIR QUALITY

Publication was supported by the Polish Ministry of Science and Higher Education

Management of Indoor Air Quality

Editor:

Marzenna R. Dudzińska
Institute of Environmental Protection Engineering,
Lublin University of Technology,
Lublin, Poland

CRC Press
Taylor & Francis Group
Boca Raton London New York Leiden

CRC Press is an imprint of the
Taylor & Francis Group, an **informa** business
A BALKEMA BOOK

CRC Press/Balkema is an imprint of the Taylor & Francis Group, an informa business

Typeset by MPS Limited, a Macmillan Company, Chennai, India

Printed and bound in Great Britain by Antony Rowe (A CPI-group Company), Eastbourne, East Sussex

Published by: CRC Press/Balkema
P.O. Box 447, 2300 AK Leiden, The Netherlands
e-mail: Pub.NL@taylorandfrancis.com
www.crcpress.com – www.taylorandfrancis.co.uk – www.balkema.nl

ISBN: 978-0-415-67266-5 (Hbk)
ISBN: 978-0-203-62762-4 (eBook)

Management of Indoor Air Quality – Dudzińska (ed)
© 2011 Taylor & Francis Group, London, ISBN 978-0-415-67266-5

Table of Contents

Management of Indoor Air Quality – Dudzińska (ed)
© 2011 Taylor & Francis Group, London, ISBN 978-0-415-67266-5

Preface

Due to efforts undertaken to minimize pollutant emissions to the atmosphere, the quality of outdoor air has increased significantly in recent decades, especially in the developed countries. But, at the same time one can observe changes in societal patterns and lifestyle. For example, it is well-known that the modern day working population spends most of its time indoors. This means not only houses and office premises, but shopping malls, fitness clubs, restaurants and transport vehicles. Recent job openings offer more "office jobs" in the service sector than in factories, where workers exposed to harmful conditions have to wear more individual protective equipment than suits and ties. With this rise in the standard of living most indoor spaces are equipped with mechanical ventilation or air-condition systems. Thus, there is a case for considering these "new" working milieus as having their own sets of environmental hazards and treating them accordingly by setting up corresponding regulations.

A number of indoor parameters (thermal comfort, odors, odorless compounds) may have a negative impact on our health and well being. Further, people often complain about the quality of the indoor environment, as they suffer from eye irritation, allergies, pneumonia, dermatosis and other diseases as well as from uncomfortable conditions. These may have an impact not only on their health but on work efficiency. These symptoms have even been recognized collectively as *Sick Building Syndrome* and widely described in the literature.

According to WHO (The Right to Healthy Indoor Air, 2000), in the developed countries, indoor air impacts more on our health than outdoor conditions, not only due to exposure time, but because the concentrations of many pollutants are higher indoors than outdoors. Volatile Organic Compounds (VOC) and SemiVolatile Organic Compounds (SVOC) are among such substances.

VOCs are emitted from building structures and finishing materials as well as from room users themselves. Studies on VOC emission rates, sources and impact have a long history, although new methods to monitor emission and examination of new types of indoor premises such as transport vehicles are still of considerable interest to scientists worldwide.

Different families of organic compounds, such as phthalates (emitted from paints and plastics), dioxins (emitted from fire places) and PBDE (polybrominated diphenyloethers), emitted from electronic equipment and textiles, as they are used as additive flame retardants belong to SVOC. Indoor air is the main source of PBDE in the environment and concentrations of PBDE in indoor air are higher than in the outdoor atmosphere. Due to their lipophilic properties they accumulate in human tissues and may cause severe health effects.

SVOC, due to their lower volatility, are generally found sorbed on suspended particles and indoor dust than in the gaseous phase. They are generally present at only very low levels, so that their sampling entails problems, especially that they may undergo transformation during long time sampling.

All substances emitted to the indoor air may undergo transformation, caused either by physical parameters (UV, temperature) or interactions with other chemical compounds (e.g. ozone, originating from outdoor sources or laser printers). Transformation of halogenated compounds may lead to formation of more hazardous products then the initial ones.

Substances emitted to the air might be in gaseous or aerosol form. Biological pollutants are found mainly as bioarosols. In the past humans themselves were thought to be the main source of bioaerosols. But ventilation and air-condition systems, which should improve the air quality, are known to act as reservoirs of microbiological pollutants, as conditions in ventilations systems are advantageous for the multiplication of fungi and bacteria (viz the famous cases of "legionaires disease" in the seventies of the last century). Increased concentrations of aerosol and bioaerosol particles decrease the indoor air quality. Microorganisms and products of their metabolism have usually negative health effects. Identification of aerosol sources and outdoor/indoor ratios are among the most popular topics of scientific papers although work on exposure and health impact are still scare.

Nowadays one can observe growing interest not only in the impact of indoor air quality on human health, but on workplace efficiency. Temperature and humidity, air exchange rates and indoor air quality, as well as noise and light intensity may influence brain activity (directly or indirectly), memory skills, and consciousness resulting in decreased job performance.

Higher indoor pollution levels can also result from constructing buildings to be low-energy consuming. Decreasing air exchange rates e.g. by tightening of windows and doors often causes inhabitants to suffer from accumulation of pollutants. Some air-condition systems, due to energy saving demands, even supply re-circulated air thus any malfunctioning of the air cleaning filters will result in secondary pollution of indoor air.

Attempts to minimize energy consumption and supply fresh air efficiently to the areas where it is really needed lead to designing and testing new ventilation systems – such as hybrid or personal. Personal ventilation, introduced previously in aircraft and vehicle cabins has become popular in office spaces, especially in humid climates, where it allows savings in energy consumption.

Scientists have long recognized the plethora of indoor air problems and the literature concerning indoor air or indoor environment contains thousands of papers addressing all of the above mentioned issues.

Twenty six reports prepared within the European Collaborative Action have been published since 1988, under the overall title "Indoor Air Quality and its Impact on Man" (since 2000 the series appears under the title "Urban Air, Indoor Environment and Human Exposure"). Reports were devoted to either a particular pollutant problem (radon, formaldehyde, nitrogen dioxide) or to a broader scope of problems such as analytical problems, human exposure, risk assessment, or ventilation and the rational use of energy.

But there are still problems with permitted levels. Some pollutants, such as formaldehyde are regulated in European and other countries. For some pollutants there are guidelines for newly built houses. Regulation systems and scope vary from country to country.

The World Health Organization document "The Right to Healthy Indoor Air" was published in 2000 and the document "Indoor air quality guidelines for selected pollutants", is now in preparation. This will include: formaldehyde, benzene, naphthalene, nitrogen dioxide, carbon monoxide, radon, trichloroethylene, tetrachloroethylene and Polycyclic Aromatic Hydrocarbons (PAH) (including benzo-a-pyrene). However, our knowledge still remains patchy. For instance, we still do not know the indoor air-related factors responsible for health problems, e.g. Sick Building Syndrome (SBS), asthma and allergies.

Further, a number of other questions still remain open:

1) how to measure indoor air quality and how to do this quickly, cheaply and reliably?
2) how to improve indoor air quality with minimum energy consumption?
3) how to minimize the secondary pollution from the air supply systems?
4) what type of pollutant we should we focus on? SVOC? Or intermediate products of indoor air reactions?
5) how are we exposed to 'modern' chemicals: plasticizers, flame retardants, pesticides, etc. and what is their real impact on our health?
6) how to establish a good model of IAQ which might be applied in practice?

Concern about the indoor air environment is different in different countries. For some, like Poland, indoor air quality is still not a primary area of interest for scientists, decision makers or citizens.

People spend a lot of money and effort to decorate their homes, but esthetic issues and costs are those mainly considered. They do not pay sufficient attention to the quality of construction materials with regard to emission and health impact.

Development of research on atmospheric pollution or water quality followed big national monitoring programs. There is a lack of such programs addressing indoor air quality problems in Poland. Limited research was performed at local level only, due to possibilities of small scientific groups. As a result we have no knowledge about the real problem, we have no data on indoor air quality in our schools, offices, and living premises, and we are not able to undertake any action to improve the situation, if needed.

Indoor air quality in Poland requires both more research and more attention from administrators and decision makers. The main needs have to be more precisely described and national multi-disciplinary research programs proposed and implemented.

From an environmental standpoint, Poland is not a unique country and its problems are similar to those of other countries in the region. This volume presents collection of 14 peer reviewed papers in Indoor Environment Engineering coming out from presentations and discussions at the workshop on "Management of Indoor Air Quality", gathering fifty engineers, chemists and biologists from Poland, Denmark and Switzerland. It was held in Kazimierz Dolny, S-E Poland in September, 2010.

Lublin, February 2011

Marzenna R. Dudzińska

Management of Indoor Air Quality – Dudzińska (ed)
© 2011 Taylor & Francis Group, London, ISBN 978-0-415-67266-5

Biohazards in the indoor environment – a role for threshold limit values in exposure assessment

R.L. Górny, M. Cyprowski, A. Ławniczek-Wałczyk & M. Gołofit-Szymczak
Biohazard Laboratory, Department of Chemical, Aerosol, and Biological Hazards, Central Institute for Labour Protection – National Research Institute, Warsaw, Poland

L. Zapór
Laboratory of Toxicology, Department of Chemical, Aerosol, and Biological Hazards, Central Institute for Labour Protection – National Research Institute, Warsaw, Poland

ABSTRACT: The hygienic evaluation of indoor environments should be based on a proper strategy followed by adequate assessment of biohazards. This paper describes a strategy applied for elaboration of hygienic standards for biological aerosols, discusses limitations of the available recommendations as well as commenting on decision making and data interpretation issues. The health- and environmentally-based approach applied for elaboration of threshold/reference limit values for bioagents are presented and discussed. Special emphasis is placed on the role of such standards in indoor air protection and the existing worldwide and Polish standards for bioaerosols are also thoroughly characterized in the paper.

Keywords: biohazards, biological aerosol, hygienic standards, threshold limit values, elaboration strategy

1 INDOOR AIR CONTAMINATION

Every biological agent, whose presence in the indoor environment is undesirable, can be recognized as its contamination. Although no premises (excluding those specially designed in pharmaceutical and biotechnological industries) are sterile and free from contamination, the presence of many biological agents in low concentrations can be recognized as "normal". The problem of pollution can appear when the level of biological contaminants increases above a certain limit recognized as normal for its background. This kind of situation can be often found when microbial indoor contamination is investigated.

Nowadays in modern societies people spend most of their time indoors and the quality of the air there is therefore an important determinant of both health and well-being. According to the "Targets 10 and 13" of the World Health Organization (WHO) document "Health21", by the year 2015, people "should live in a safer physical environment, with exposure to contaminants hazardous to health at levels not exceeding internationally agreed standards" (WHO 1998). Based on these assumptions, WHO in the report "The Right to Healthy Indoor Air" stated that everyone has the right "to breathe healthy indoor air, to adequate information about potentially harmful exposures, and to be provided with effective means for controlling at least part of their indoor exposures" (WHO 2000). In that light, the control of biohazards in indoor environment often seems to be inadequate. In the perception of society, the control of physical or chemical hazards is obvious, required, and frequently performed. Regarding biohazards, though the public has become more aware of their presence indoors, their control is still relatively rare. Even if performed, a lack of widely accepted quantitative guidelines or thresholds for acceptable levels of microbial contamination prevents adequate assessment of the hygienic quality of indoor environments.

2 HARMFUL MICROBIAL AGENTS IN INDOOR ENVIRONMENT

According to the definition, among harmful biological agents are: microorganisms, including those which have been genetically modified, cell cultures, including (growing in vitro) cells derived from multicellular organisms, and human endoparasites, which are able to provoke infection, allergy or toxic reaction, directly or through the structures and substances produced by them, as well as cellular or non-cellular organisms capable of replication or transferring a genetic material. The cited definition includes a wide variety of agents. Due to the generality of their effects in the indoor environment, the group of harmful agents of microbial origin (HMA) has been selected for discussion below.

Through the ages, the development of civilization has been integrally connected with microbial expansion to places inhabited by humans. HMA have the ability to grow and colonize both natural and artificial materials of organic as well as inorganic origin. Causing their biodegradation and biodeterioration, they deliver many biologically active structures and substances into the air posing potential threats to the exposed individuals. In an indoor environment, the airborne transport of HMA is the most common route for their dispersion and usually an exposure to high concentrations of bioaerosols can cause numerous immunopathogenic reactions. Among HMA there are:

- agents that cause infective and invasive diseases (e.g. viruses, bacteria, fungi),
- toxins and compounds with similar toxic effects (e.g. bacterial exotoxins and endotoxins, fungal mycotoxins, glucans and volatile organic compounds),
- carcinogens (e.g. aflatoxins produced by some species of *Aspergillus* genera or trichothecenes produced by some species of *Fusarium* genera),
- allergens (of bacterial and fungal origin),
- fragments (fine, i.e. submicrometric and nanometric particulates of bacterial and fungal origin).

In the Directive 2000/54/EC and in an Ordinance of the Polish Ministry of Health (2005), biological agents are classified into four groups according to their level of risk of infection:

- group 1 – agents, that are unlikely to cause human disease;
- group 2 – agents, that can cause human disease and might be a hazard to workers; they are unlikely to spread to the community; there is usually effective prophylaxis or treatment available;
- group 3 – agents, that can cause severe human disease and present a serious hazard to workers; they may present a risk of spreading to the community, but there is usually effective prophylaxis or treatment available;
- group 4 – agents, that cause severe human diseases and are in serious hazard to workers; they may present a high risk of spreading to the community; there is usually no effective prophylaxis or treatment available.

In the indoor environment, the above mentioned HMA are present as aerosol particulates, i.e. as single cells and spores, their fragments as well as aggregates of biological particulates only or mixed aggregates consisting of biological particulates with dust particles and fibers. Their sizes vary from nanometric values (e.g. bacterial endotoxin), through submicron (e.g. bacterial and fungal fragments), up to diameters above 100 μm (e.g. plant pollens). The character of HMA airborne structures, aerodynamic diameters, shapes, densities, chemical composition and reactivity together with other parameters such as air velocity, breathing manner and lung ventilation volume (which depends on age as well as dynamics of personal activities) decide their depth of penetration into the respiratory tract, site of deposition and types of interaction with the cells of exposed individuals.

3 INDOOR BIOAEROSOLS – WHAT AND HOW TO STUDY

Bioaerosols can consist of: viable microbial particles which are able to produce colonies on proper medium, viable particles unable to grow and produce such structures as well as non-viable microorganisms or fragments derived from the microbial cells. So far, there is a lack of a single universal sampling method that allows analysts to efficiently collect all these four types of airborne particulates in a fixed volume of air (physical sampling efficiency) while simultaneously preserving all their biological features (biological sampling efficiency). Thus, in everyday practice, the assessment of microbial air quality is limited to the measurement of viable bacterial and fungal particles, i.e. single microbial cells, spores or their aggregates which are able to grow on proper agar medium as separate colonies. Despite the limitations concerning sampling efficiency procedures, the current state of knowledge allows collection of bioaerosol particles using volumetric methods, which are based on active sampling of an appropriate volume of the air containing intact microorganisms or their fragments (Dutkiewicz et al. 2007, Górny 2010).

At present, nowhere in the world are there distinct regulations to standardize microbial air sampling in indoor environments. Hence, the regulations contained in four Polish Standards describing the methods of workplace control can be (in part, at least) applied for this purpose. A brief overview of these norms is given below. The Polish Standard PN-EN 13098 accepted by the Polish Committee for Standardization (PKN) in 2002 (and replaced in 2007) entitled "Workplace atmosphere – Guidelines for measurement of airborne microorganisms and endotoxin" contains the basic definitions, gives recommendations regarding the use of volumetric sampling methods, legitimizing the possible application of the analysis of microbial cell compounds (such as endotoxins or glucans) as well as the primary (ATP) and secondary (mycotoxin) metabolites for the assessment of microbial air contaminants. The next norm accepted by the PKN in 2002 (and replaced in 2006) is the Polish Standard PN-EN 14031 "Workplace atmospheres – Determination of airborne endotoxins", in which the guideline for exposure assessment of airborne endotoxins at workplaces is given. The methods of sampling, transport, storage and endotoxin analysis techniques are broadly described. In the same year, the PKN accepted (and in 2010 replaced) the European Standard "Workplace atmospheres – Guide for the application and use of procedures for the assessment of exposure to chemical and biological agents" as the Polish Standard PN-EN 14042. This standard contains guidelines on how to choose measurement procedures, select sampling devices for appropriate applications, use and service samplers applied for sampling of chemical and biological agents at workplaces located in both outdoor and non-industrial indoor environments. It describes several sampling methods and practical rules for using them in personal and stationary sampling. Regarding biological agents, it describes the methods of airborne microorganisms and endotoxin sampling using impactors, impingers and filter samplers. The latest Polish Standard PN-EN 14583 accepted by PKN in 2005 (and replaced in 2008) "Workplace atmospheres – Volumetric bioaerosol sampling devices – Requirements and test methods" defines the requirements for testing the efficiency of volumetric samplers used for exposure assessment of bioaerosols at workplaces.

4 HYGIENIC STANDARDS FOR BIOAEROSOLS – ELABORATION STRATEGY

In contrast to many chemical and physical hazards, on the global scale there is a lack of acceptable hygienic standards for harmful biological agents. The main reason for this is the lack of well-documented dose-response relationships, i.e. epidemiologically proven connections between the exposure to specific HMA and adverse health effects caused by their exact dose(s). Moreover, the sensitivity of each organism is individual and the strength of immunological reaction to the specific HMA is usually not the same. Despite progress in development of aerosol sampling techniques and analytical methods over the last two decades, the worldwide scientific database on bioaerosols is still insufficient to quantitatively and qualitatively characterize HMA precisely in the environment. If the threshold, limit or reference values are established, they are usually connected with the clinical picture of the specific disease caused by HMA, taking into consideration the presence of the bioagent in a certain element of the environment only. Nevertheless, in contravention of these limitations, the reference values expressed in numbers are applied to facilitate interpretation of the measurement data. They are usually used as arbitral or comparative values. Table 1 summarizes the most important features of both of them.

The strategy of elaboration of hygienic standards for HMA takes into consideration many factors. The sampling method used and several environmental, source, quantitative and qualitative criteria play a key role in such a process. Their interdependence is presented graphically in Figure 1.

Table 1. The features of numerical reference values for Harmful Microbiological Agents (HMA).

Numerical reference values	
Arbitrary	Relative/comparative
1. Concentration of HMA, which is acceptable or unacceptable 2. Usually for total microbial concentration, group of bioagents or given species 3. Proposed by individual researchers, expert groups or as a result of cross-sectional/large-scale research studies 4. Not related to the specific health outcome (does not precisely reflect dose-response relationship) 5. Often equal or close to detection level of particular analytical method	1. Usually based on simultaneous measurements of HMA in indoor and outdoor environment 2. If the indoor/outdoor ratio is <1, it is interpreted as a "lack of contamination" or acceptable level of contamination 3. Identification of indoor sources of HMA 4. Suitable for qualitative or frequency of appearance comparisons

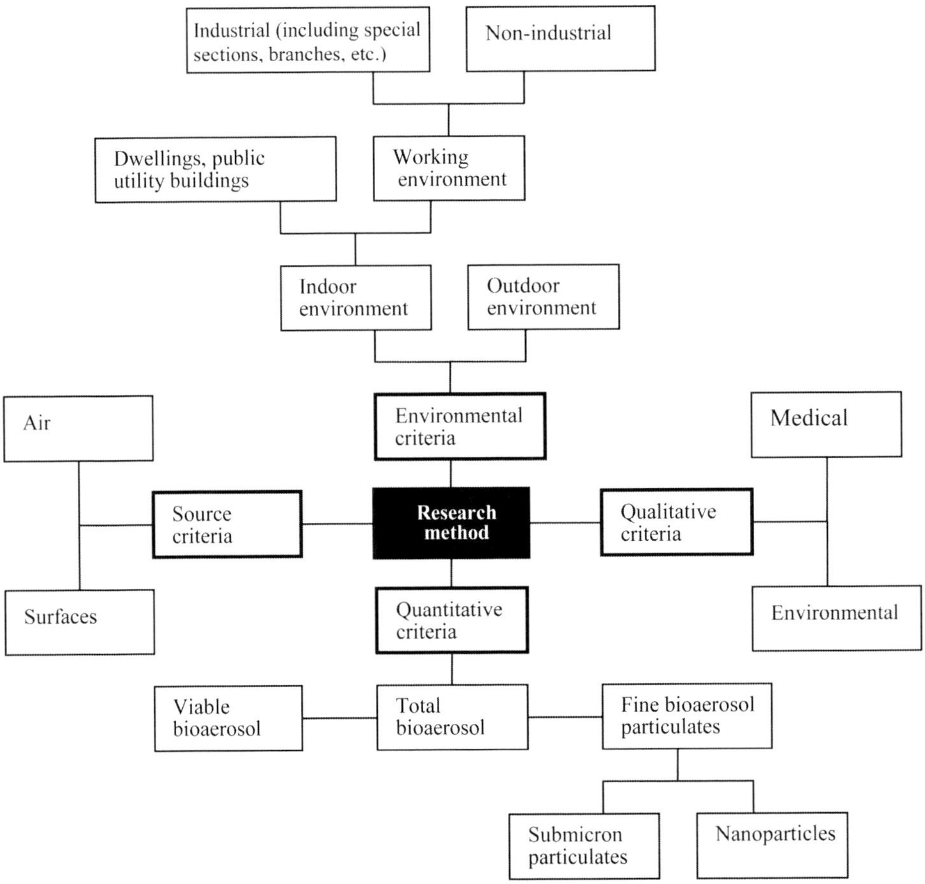

Figure 1. Strategy for creation of hygienic standards for harmful microbiological agents (Górny 2010).

5 CLINICAL CRITERIA VERSUS "ENVIRONMENTAL PHILOSOPHY"

Airborne microorganisms can be responsible for many adverse health effects from allergic reactions (e.g. asthma, hypersensitivity pneumonitis, allergic rhinitis), through infections (caused by microbial growth on or in the host organism, e.g. aspergillosis), to toxic reactions (connected mainly with toxins, secondary metabolites or microbial cell wall constituents) and other non-specific symptoms referred to as "sick building syndrome", "tight building syndrome" or "building related illness". From the point of view of clinical outcomes caused by exposure to HMA, the ideal situation would be if: a) a hygienic standard would be created based on the relationship between exposure (characterized by the type and concentration of HMA) and health effects they caused; b) such a relationship would be epidemiologically proven; c) experimental evidence proving this relationship would become known; d) such a relationship would be clinically proved also, i.e. medical evidence that a specific agent was responsible for the observed specific health effect would exist and e) for such a relationship, quantity and taxonomic origin of HMA responsible for the observed adverse health effect would be known. As we know, there is practically no HMA (except bacterial endotoxin for which we are close to defining a dose-response relationship) for which the creation of hygienic standard based on clinical criteria would be possible. Thus, the effective solution in this situation could be so-called "environmental philosophy" on hygienic standard building. In brief, in the situation in which a solid link between the concentration of investigated parameters and resulting adverse health effect cannot be effectively established, then, based on the multiple biological agent concentration measurements, the reference values should enable an evaluation of the quality of environment, as well as the determination of "what is typical or acceptable" and "what is atypical or inacceptable" for a specific type of environment (or its fragment). At present, due to the lack of possibilities to determine the relationship between the dose of HMA and health outcome in a precise way, the creation of a hygienic standard based on the "environmental philosophy" seems to be a reasonable alternative for the "clinical" approach.

6 WORLWIDE AND POLISH PROPOSALS OF HYGIENIC STANDARDS FOR BIOAEROSOLS IN INDOOR ENVIRONMENT

The hygienic standard values available in the scientific literature for bioaerosols in industrial and non-industrial indoor environments with appropriate comments are presented in Tables 2–8.

Their analysis enables users to determine the ranges of concentration values, which have been proposed by organizations, institutions, national expert committees, independent groups of researchers or individual scientists within the last 140 years. All these values of standards, norms and proposals for threshold limits, which help to interpret empirically obtained data, have either arbitral or comparative character. A wide review of these proposals, including those for numerous HMA such as bacteria, fungi, microbial related compounds as well as for special environments and those in which high air quality is required can be found in the publications by Brandys and Brandys (2007) or Górny (2004). Regarding particular HMA, so far proposed values are as follows ("cfu" means "colony forming units", i.e. the number of microbial cells, spores or their aggregates that are able to grow on proper agar medium as separate colonies):

- for total number of bacteria: $\leq 1.0 \times 10^3 \div \leq 7.0 \times 10^3$ cfu/m^3 for residential rooms and non-industrial workplaces and $\leq 7.5 \times 10^2 \div 1.0 \times 10^7$ cfu/m^3 for productive and industrial premises; for the pathogenic microorganisms, there is no safety level in the above mentioned premises (the threshold limit should be 0 cfu/m^3),
- for Gram-negative bacteria: $1.0 \times 10^3 \div 2.0 \times 10^4$ cfu/m^3 for productive and industrial premises,
- for bacterial endotoxin: $0.005 \div 0.2$ μg/m^3 for productive and industrial premises,
- for fungi: $1.0 \times 101 \div 1.0 \times 104$ cfu/m^3 for residential rooms and non-industrial workplaces and $< 1.0 \times 102 \div 1.0 \times 107$ cfu/m^3 for productive and industrial premises, for the pathogenic microorganisms, there is no safety level in the above mentioned premises (the threshold limit should be 0 cfu/m^3),
- for subtilisin: 0.06 μg/m^3 for productive and industrial premises.

Table 2. Hygienic standards for bacteria in indoor environment.
a) Proposals of governmental organizations/institutions for total number of bacteria:

Year of publishing	Organization/ institution or country	Source	Hygienic standard value/reference limit value
1986	AIHA (American Industrial Hygiene Association)	Biohazards Reference Manual	0 cfu/m^3 – no safety level for pathogenic microorganisms
1989	DECOS Netherlands	Research Methods in Biological Indoor Air Pollution	$>10^3$ cfu/m^3 (for Gram-positive bacteria) – health hazard
1993	CEC (Commission of the European Communities)	Report No. 12: Biological Particles in Indoor Environment	$<1.0 \times 10^2$ cfu/m^3 (homes) and $<5.0 \times 10^1$ cfu/m^3 (non-industrial rooms) – very low contamination $<5.0 \times 10^2$ cfu/m^3 (homes) and $<1.0 \times 10^2$ cfu/m^3 (non-industrial rooms) – low contamination $<2.5 \times 10^3$ cfu/m^3 (homes) and $<5.0 \times 10^2$ cfu/m^3 (non-industrial rooms) – average/elevated contamination $<1.0 \times 10^4$ cfu/m^3 (homes) and $<2.0 \times 10^3$ cfu/m^3 (non-industrial rooms) – high contamination $>1.0 \times 10^4$ cfu/m^3 (homes) and $>2.0 \times 10^3$ cfu/m^3 (non-industrial rooms) – very high contamination
1996	China: Ministry of Health	Chinese National Standards	$\leq 1.0 \times 10^3$–2.5×10^3 cfu/m^3 – limit value for hotels (GB9663–1996) $\leq 2.5 \times 10^3$–4.0×10^3 cfu/m^3 – limit value for public places of entertainment (GB9664–1996) $\leq 4.0 \times 10^3$ cfu/m^3 – limit value for barber and beauty shops (GB9666–1996) $\leq 4.0 \times 10^3$ cfu/m^3 – limit value for swimming pools (GB9667–1996) $\leq 4.0 \times 10^3$ cfu/m^3 – limit value for school classes (GB9668–1996) $\leq 2.5 \times 10^3$ cfu/m^3 – limit value for libraries, museums, galleries (GB9669–1996) $\leq 7.0 \times 10^3$ cfu/m^3 – limit value for exhibition halls (GB9669–1996) $\leq 7.0 \times 10^3$ cfu/m^3 – limit value for shopping centers, bookstores (GB9670–1996) $\leq 7.0 \times 10^3$ cfu/m^3 – limit value for waiting rooms at railway stations, bus stations, harbors (GB9672–1996) $\leq 4.0 \times 10^3$ cfu/m^3 – limit value for waiting rooms at airports (GB9672–1996) $\leq 4.0 \times 10^3$ cfu/m^3 – limit value for trains, ships (GB9673–1996) $\leq 2.5 \times 10^3$ cfu/m^3 – limit value for airplanes (GB9673–1996)
2001	China: Ministry of Health	Chinese National Standards	$\leq 2.5 \times 10^3$ cfu/m^3 – limit value for residential premises (255–2001)

(*Continued*)

Table 2. (*Continued*)

b) Proposals of governmental organizations/institutions for Gram-positive bacteria:

Year of publishing	Organization/ institution or country	Source	Hygienic standard value/reference limit value
1989	DECOS Netherlands	Research Methods in Biological Indoor Air Pollution	$>10^3$ cfu/m^3 – health hazard

c) Individual proposals/groups of researchers for total number of bacteria:

Year of publishing	Author (authors)	Hygienic standard value/reference limit value
1894	Bujwid	50/liter – acceptable value for the air in dwellings
1941	Bourdillon et al.	1.765×10^3 cfu/m^3 – maximum admissible concentration (for sum of bacteria and fungi)
1955	Topley & Wilson	2.67×10^3 (for sum of bacteria and fungi) and 3.0×10^1 (for *Streptococcus*) cfu/m^3 – maximum admissible concentration in shops 2.825×10^3 (for sum of bacteria and fungi) and 4.5×10^1 (for *Streptococcus*) cfu/m^3 – maximum admissible concentration in schools
1955	Wells	1.5×10^3 (summer) and 4.5×10^3 (winter) (for sum of bacteria and fungi) and 1.6×10^1 (summer) and 3.6×10^1 (winter) (for *Streptococcus*) cfu/m^3 – clean air 2.5×10^3 (summer) and 7.0×10^3 (winter) (for sum of bacteria and fungi) and 3.6×10^1 (summer) and 1.24×10^2 (winter) (for *Streptococcus*) cfu/m^3 – contaminated air
1985	Clark	1.0×10^7 cfu/m^3 – safe concentration in the air of productive rooms
1988	Donham	4.3×10^5 cfu/m^3 – standard value (for sum of bacteria and fungi) for exposure in animal confinement buildings
1989	Nevalainen	4.5×10^3 cfu/m^3 – maximum concentration in dwellings
1989	Erman et al.	5.0×10^4 cfu/m^3 – standard value (for sum of bacteria and fungi) for exposure in productive rooms
1990	Reponen et al.	5.0×10^3 cfu/m^3 – maximum standard concentration in dwellings
1992	Malmros et al.	5.0×10^3–1.0×10^4 cfu/m^3 – threshold limit value for occupational exposure
1994	Binnie	$\leq 7.5 \times 10^2$ cfu/m^3 – contaminated air, if the concentration (bacteria and fungi together) exceeds this value and if among the identified species infectious and allergenic organisms are not present.
1995	Heida et al.	$>10^4$ cfu/m^3 – health hazard

d) Individual proposals/groups of researchers for Gram-negative bacteria:

Year of publishing	Author (authors) or organization	Hygienic standard value/reference value
1985	Clark	1.0×10^3 cfu/m^3 – safe concentration in the air of productive rooms
1990	Buyanov et al.	2.0×10^4 cfu/m^3 – standard limit value for exposure in productive rooms (for *E. coli*)
1992	Malmros et al.	1.0×10^3 cfu/m^3 – threshold limit value for occupational exposure

Table 3. Hygienic standards for bacterial endotoxin.

Year of publishing	Author (authors) or organization	Hygienic standard value/reference limit value in μg/m³ (or endotoxin units, EU/m³, if available)
1984, 1991	Haglind and Rylander; Donham	0.08
1985	Clark	0.1
1987, 1992	Rylander; Malmros, Sigsgaard & Bach	0.1–0.2
1987	Castellan, Olenchock, Kinsley & Hankinson	0.009
1988	Palchak, Cohen, Ainslie and Hoerner	0.03
1992	Smid, Heederik, Mensink, Houba, Boleij	0.015
1993	Dutkiewicz & Molocznik	0.2 (2000)
1998	DECOS Netherlands	0.0045 (5)
1999	Laitinen	0.025
2000	Donham, Cumro, Reynolds, Merchant	(100)
2002	Górny & Dutkiewicz	0.2 (2000)
2003	Douwes, Thorne, Pearce, Heederik; Martinez, Rao & Burton	(200)
2004	Madsen, Mårtensson, Schneider & Larsson	0.015
2005	Suva Switzerland	(1000)

Table 4. Hygienic standard for fungi in indoor environment.
a) Proposals of governmental organizations/institutions:

Year of publishing	Organization/ institution or country	Source	Hygienic standard value/reference limit value
1986	AIHA (American Industrial Hygiene Association)	Biohazards Reference Manual	0 cfu/m³ – no safety level for pathogenic microorganisms
1988	CMHC (Canada Mortgage and Housing Corporation)	Determination of Fungal Propagules in Indoor Air	0 cfu/m³ – no need of action $\geq 5.0 \times 10^1$ cfu/m³ – if refer to one species only, then the source of microorganism should be identify $\leq 1.5 \times 10^2$–2.0×10^2 cfu/m³ – if refer to many species, no need of action $\geq 2.0 \times 10^2$ cfu/m³ – if refer to many species, further inspection is required $\leq 4.0 \times 10^2$–5.0×10^2 cfu/m³ – if refer to *Cladosporium* and *Alternaria*, no need of action $\geq 5.0 \times 10^2$ cfu/m³ – if refer to *Cladosporium* and *Alternaria*, a reason of such state should be found
1988	WHO (World Health Organization)	IAQ: Biological Contaminants	0 cfu/m³ – in indoor environment, there shouldn't be pathogenic and toxigenic fungi $>5.0 \times 10^1$ cfu/m³ – if refer to one species only, the action should be undertaken $\leq 1.5 \times 10^2$ cfu/m³ – acceptable, if it is a "mixture" of different species $\leq 5.0 \times 10^2$ cfu/m³ – acceptable, if it is *Cladosporium* or other phytomoulds

(*Continued*)

Table 4. (*Continued*)

Year of publishing	Organization/ institution or country	Source	Hygienic standard value/reference limit value
1989	DECOS Netherlands	Research Methods in Biological Indoor Air Pollution	$>1.0 \times 10^4$ cfu/m^3 – health hazard $>5.0 \times 10^2$ cfu/m^3 – if it is one potentially pathogenic species only, health hazard
1989	AIHA (American Industrial Hygiene Association)	Practitioner's Approach to IAQ Investigations	$\geq 1.0 \times 10^4$ cfu/m^3 – it suggests not typical situation High indoor/outdoor concentration ratio indicates indoor source of emission
1989	ACGIH (American Conference of Governmental Industrial Hygienists)	Guidelines for the Assessment of Bioaerosols	$<1.0 \times 10^2$ cfu/m^3 – not contaminated environment Indoor/outdoor concentration ratio <1 – not contaminated environment, if the same genera/species are present
1991	Bowser Technical Inc.	Testing of Older Houses for Microbiological Pollutants	$\leq 2.0 \times 10^2$ cfu/m^3 – recommended comparison of indoor/outdoor concentrations $>2.0 \times 10^2$ cfu/m^3 – if species different than *Alternaria* and *Cladosporium*, the action should be undertaken $>5.0 \times 10^2$ cfu/m^3 – if among different species are *Alternaria* and *Cladosporium*, the action should be undertaken
1991	Nordic Council	Criteria Documents from the Export Group	1.0×10^1–1.0×10^4 cfu/m^3 – typical level for "sick building" and outdoor air
1992	US OSHA (United States Occupational Safety and Health Administration)	Technical Manual	$>1.0 \times 10^3$ cfu/m^3 – contamination index
1993	CEC (Commission of the European Communities)	Report No. 12: Biological Particles in Indoor Environment	$<5.0 \times 10^1$ cfu/m^3 (homes) and $<2.5 \times 10^1$ cfu/m^3 (non-industrial rooms) – very low contamination $<2.0 \times 10^2$ cfu/m^3 (homes) and $<1.0 \times 10^2$ cfu/m^3 (non-industrial rooms) – low contamination $<1.0 \times 10^3$ cfu/m^3 (homes) and $<5.0 \times 10^2$ cfu/m^3 (non-industrial rooms) – average/elevated contamination $<1.0 \times 10^4$ cfu/m^3 (homes) and $<2.0 \times 10^3$ cfu/m^3 (non-industrial rooms) – high contamination $>1.0 \times 10^4$ cfu/m^3 (homes) and $>2.0 \times 10^3$ cfu/m^3 (non-industrial rooms) – very high contamination

(*Continued*)

Table 4. (*Continued*)

Year of publishing	Organization/ institution or country	Source	Hygienic standard value/reference limit value
1993	Russian Federation: State Committee for Hygiene and Epidemiological Surveillance	Standard: Maximum Allowable Concentrations of Harmful Substances	1.0×10^3–1.0×10^4 cells/m^3 – depending on the species
1994	Cutter Information Corp.	IAQ Update: Biocontaminants in Indoor Environments	3.0×10^2 cfu/m^3 – maximum admissible concentration for common fungi (e.g. *Cladosporium*) 2.0×10^2 cfu/m^3 – maximum admissible concentration for total fungal flora 1.5×10^2 cfu/m^3 – maximum admissible concentration for species other than pathogenic and toxicogenic 1.0×10^2 cfu/m^3 – maximum admissible concentration, if there is no people with lower immunodeficiency
1995	ACGIH (American Conference of Governmental Industrial Hygienists)	Air Sampling Instruments for Evaluation of Atmospheric Contaminants	$<1.0 \times 10^2$ cfu/m^3 – low, for rooms where high cleanliness is required and hospitals 1.0×10^2–1.0×10^3 cfu/m^3 – average, for other indoor rooms and outdoor air $>1.0 \times 10^3$ cfu/m^3 – high, when animal production
1995	Indoor Air Quality Association Inc.	IAQ Standard No. 95–1 Recommended for Florida	$<3.0 \times 10^2$ cfu/m^3 – for commonly present fungi $<1.5 \times 10^2$ cfu/m^3 – for a mixture of other fungi than pathogenic and toxicogenic
1995	New York City: Department of Health	Guidelines on Assessment and Remediation of *S. atra* in Indoor Environments	10^3–10^4 cfu/m^3 – immediate evacuation of exposed people
1997	Finnish Institute of Occupational Health	Technical Report	$<1.0 \times 10^2$ cfu/m^3 – "acceptable" contamination level 1.0×10^2–5.0×10^2 cfu/m^3 – contamination level with tes-ting required $>5.0 \times 10^2$ cfu/m^3 – "unacceptable" contamination level
2003	Institut fur Baubiologie & Oekologie, Germany	Standard SBM–2003	$<2.0 \times 10^2$ cfu/m^3 – no anomaly 2.0×10^2–5.0×10^2 cfu/m^3 – weak anomaly 5.0×10^2–1.0×10^3 cfu/m^3 – strong anomaly $>1.0 \times 10^3$ cfu/m^3 – extreme anomaly

b) Individual proposals/group of researchers:

Year of publishing	Author (authors)	Hygienic standard value/reference value
1941	Bourdillon et al.	1.765×10^3 cfu/m^3 – maximum admissible concentration (for sum of bacteria and fungi)
1955	Topley & Wilson	2.67×10^3 cfu/m^3 – maximum admissible concentration (for sum of bacteria and fungi) in shops 2.825×10^3 cfu/m^3 – maximum admissible concentration (for sum of bacteria and fungi) in schools

(*Continued*)

Table 4. (*Continued*)

Year of publishing	Author (authors)	Hygienic standard value/reference value
1955	Wells	1.5×10^3 (summer) and 4.5×10^3 (winter) cfu/m^3 – clean air 2.5×10^3 (summer) and 7.0×10^3 (winter) cfu/m^3 – contaminated air
1979	Berk et al.	2.0×10^1–7.0×10^2 cfu/m^3– no negative health outcomes
1984	Holmberg	$<2.2 \times 10^3$ cfu/m^3– environment free from contamination 1.0×10^4–1.5×10^4 cfu/m^3 – certain growth on surfaces
1984	Morey et al.	$>1.0 \times 10^3$ cfu/m^3 – the action should be undertaken
1984	Solomon & Burge	$<1.6 \times 10^3$ cfu/m^3 – not contaminated residential indoor air $<6.0 \times 10^3$ cfu/m^3– maximum level of contamination
1985	Clark	1.0×10^7 cfu/m^3 – safety level in the air of productive rooms
1987	Ohgke et al.	$>1.0 \times 10^2$ cfu/m^3– indoor source of contamination, further testing is required
1988	Lacey & Crook	1.0×10^3–1.0×10^4cfu/m^3 – not contaminated air
1988	Donham	4.3×10^5 cfu/m^3 – standard value (for sum of bacteria and fungi) for exposure in animal confinement buildings
1988	Miller et al.	0 cfu/m^3 – in indoor environment, there shouldn't be pathogenic and toxicogenic fungi $\geq 5.0 \times 10^1$ cfu/m^3 – not contaminated air, if one fungal species was identified only $\leq 1.5 \times 10^2$ cfu/m^3 – not contaminated air, if several fungal species were identified $\leq 3.0 \times 10^2$ cfu/m^3 – not contaminated air, if the fungi represent commonly present phytomolds
1989	Erman et al.	5.0×10^4 cfu/m^3 – standard value (for sum of bacteria and fungi) for exposure in productive rooms
1990	Burge	If the indoor microbial aerosol differs qualitatively from outdoor bioaerosol, its concentration (constantly) at least two times exceeds outdoor concentration and is higher than 1.0×10^3 cfu/m^3 – hygienic action should be undertaken
1990	Reponen et al.	$>5.0 \times 10^2$ cfu/m^3 – additional indoor source of emission exists (in winter) Indoor/outdoor concentration ratio >1 – additional indoor source of emission exists (in summer)
1990	Reynolds et al.	$>5.0 \times 10^2$ cfu/m^3 – contaminated air Significant difference of indoor/outdoor concentration – additional indoor source of emission exists
1991	Godish	$<1.0 \times 10^2$ cfu/m^3 – environment free from contaminants $>1.0 \times 10^3$ cfu/m^3 – high level of contamination 1.0×10^2–1.0×10^3cfu/m^3 – self-interpretation and assessment of environment cleanliness by person responsible for the testing
1991	Malmberg	$\geq 5.0 \times 10^1$ cfu/m^3 – if the fungi represent the one species only, further testing is required $\leq 1.5 \times 10^2$ cfu/m^3 – not contaminated air, if the fungi represent many species $\leq 5.0 \times 10^2$ cfu/m^3 – not contaminated air (in summer), if the fungi are from the group of common tree and leaf molds
1992	Malmros et al.	5.0×10^3–1.0×10^4 cfu/m^3 – threshold limit value for occupational exposure
1993	Yang et al.	2.0×10^2 cfu/m^3 – maximum level of contamination
1994	Binnie	$\leq 7.5 \times 10^2$ cfu/m^3 – not contaminated air, if the sum of bacterial and fungal concentrations doesn't exceed this value and no infective and allergenic organisms were present

(*Continued*)

Table 4. (*Continued*)

Year of publishing	Author (authors)	Hygienic standard value/reference value
1995	Heida et al.	$>10^4$ cfu/m^3 – health hazard 5.0×10^2 cfu/m^3 – threshold limit value (for single fungal species as *Aspergillus fumigatus*)
1999	Klanova & Drahonovska	$>2.0 \times 10^3$ cfu/m^3 – serious risk factor for health
2005	Clark	$<2.5 \times 10^2$ cfu/m^3 – natural background level 2.5×10^2–1.0×10^3 cfu/m^3 – possible contamination source(s) $>1.0 \times 10^3$ cfu/m^3 – likely contamination source(s)

Table 5. Microbial cleanliness of the air in hospital rooms.

Year of publishing	Author (authors), organization/ institution/company or country	Hygienic standard value/reference value
1955	Topley and Wilson	7.0×10^1 (for sum of bacteria and fungi) and 0 (for *Streptococcus*) cfu/m^3 – maximum admissible concentration in neurosurgical operating rooms 2.5×10^2 (for sum of bacteria and fungi) and 0 (for *Streptococcus*) cfu/m^3 – maximum admissible concentration in surgical operating rooms 7.0×10^2 (for sum of bacteria and fungi) and 0 (for *Streptococcus*) cfu/m^3 – maximum admissible concentration in treatment and dressing rooms
1976	American Academy of Orthopedic Surgeons	3.5×10^1 cfu/m^3 – maximal number of microorganisms in room of class "1" 1.75×10^2 cfu/m^3 – maximal number of microorganisms in room of class "5" 7.0×10^2 cfu/m^3 – maximal number of microorganisms in room of class "20"
1981	AFNOR France: standard NF S 90-351	5.0×10^0 cfu/m^3 – bacterial class of cleanliness in empty operating room "B 5" 2.0×10^1 cfu/m^3 – bacteria class of cleanliness in empty operating room "B 20" 1.0×10^2 cfu/m^3 – bacterial class of cleanliness in empty operating room "B 100"
1984	Kruczkowski	7.0×10^1 cfu/m^3 – maximal number of bacteria in room of class "I" 3.0×10^2 cfu/m^3 – maximal number of bacteria in room of class "II" 7.0×10^2 cfu/m^3 – maximal number of bacteria in room of class "III"
1987	Schweizerische Krankenhausinstitut (SKI)	$<1.0 \times 10^1$ cfu/m^3 – maximal number of microorganisms in room of class "I" 5.0×10^1 cfu/m^3 – maximal number of microorganisms in room of class "IIb" 2.0×10^2 cfu/m^3 – maximal number of microorganisms in room of class "II" 5.0×10^2 cfu/m^3 – maximal number of microorganisms in room of class "III"

(*Continued*)

Table 5. (*Continued*)

Year of publishing	Author (authors), organization/ institution/company or country	Hygienic standard value/reference value
1995	Italian Institute of Health	1.0×10^1 cfu/m^3 – maximal number of microorganisms in room of class "A" 2.0×10^2 cfu/m^3 – maximal number of microorganisms in room of class "B" 5.0×10^2 cfu/m^3 – maximal number of microorganisms in room of class "C"
1995	Whyte & Coll	$<2.0 \times 10^1$ and $<1.0 \times 10^1$ cfu/m^3 – admissible number of microbial colonies in operating field when surgeons are "normally" and "hermetically" dressed, respectively (ultra clean operating room, air-conditioned with the sterilized air) $<1.0 \times 10^1$ and $<1.0 \times 10^0$ cfu/m^3 – admissible number of microbial colonies in a distance of $\leq$30 cm from operating wound when surgeons are "normally" and "hermetically" dressed, respectively (ultra clean operating room, air-conditioned with the sterilized air) $<5.0 \times 10^0$ cfu/m^3 – admissible number of microbial colonies near the HEPA filter at the outlet from the room
1995	ACGIH (American Conference of Governmental Industrial Hygienists)	$<1.0 \times 10^2$ cfu/m^3 – low, for rooms where higher cleanliness is required and for hospitals
1995	Rabino	$0–2.5 \times 10^1$ cfu/m^3 – assessment of microbial cleanliness: very good $2.6 \times 10^1–1.25 \times 10^2$ cfu/m^3 – assessment of microbial cleanliness: good $12.6 \times 10^2–2.5 \times 10^2$ cfu/m^3 – assessment of microbial cleanliness: weak $2.51 \times 10^2–3.75 \times 10^2$ cfu/m^3 – assessment of microbial cleanliness: bad $>3.75 \times 10^2$ cfu/m^3 – assessment of microbial cleanliness: very bad
1995	Sartorius	1.0×10^0 cfu/m^3 – cleanliness class of rooms and sterile zones "MCB–1" 1.8×10^1 cfu/m^3 – cleanliness class of rooms and sterile zones "MCB–2" 8.8×10^1 cfu/m^3 – cleanliness class of rooms and sterile zones "MCB–3"
1996	China: Ministry of Health	$\leq 4.0 \times 10^3$ cfu/m^3 – limit value for hospital waiting rooms (GB9671–1996)
1999 (replaced in 2008)	Germany: standard DIN 1946-4	4.0×10^0 cfu/m^3 – acceptable concentration of microorganisms 1.0×10^1 cfu/m^3 – limit concentration of microorganisms
2003	Kalliokoski	$<1.0 \times 10^1$ cfu/m^3 – normal risk level 1.0×10^2 cfu/m^3 – high risk level

Table 6. The limits recommended in monitoring of microbial contaminants of cleanrooms in action.

a) according to Ordinance 1882 of the Polish Ministry of Health on requirements of Good Manufacturing Practice and European Commission Guide to Good Manufacturing Practice (EC GMP): Manufacturing of Sterile Medicinal Products (Rev. to Annex 1):

Class	Recommended limits of microbial contamination (mean values) Air sample (cfu/m^3)
A	$<1.0 \times 10^0$
B	1.0×10^1
C	1.0×10^2
D	2.0×10^2

b) according to the American Pharmacopoeia (USP, chapter 1116, US Fed. Std. 209E):

Class according to US Fed. Std. 209E	Equivalent of proper class according to EC GMP	Recommended limits of microbial contamination (mean values) Air sample (cfu/m^3)
100	A	$<3.0 \times 10^0$
10000	B	$<2.0 \times 10^1$
100000	C	$<1.0 \times 10^2$
is not applied	D	–

c) Standard of NASA – NHB 5340.2 (NASA Standard for Clean Room and Work Stations (1967):

Class according to US Fed. Std. 209E	Concentration of airborne microorganisms #/ft^3*	Concentration of settled microorganisms during one week on the surface of 1 ft^2
100	1.0×10^{-1}	1.2×10^3
10000	5.0×10^{-1}	6.5×10^4
100000	2.5×10^0	3.0×10^5

*ft – foot

d) Proposal of standard for cleanrooms according to International Committee of Contamination Control Societies – ICCCS (1978):

Class	Microbial concentration in 1 m^3 of the air
I	9.0×10^1
II	1.7×10^1
III	3.0×10^0
IV	1.0×10^0

In the special environments like hospital rooms or cleanrooms in action, the threshold limit values for airborne microbial contaminants should be within the range of $1.0 \times 10^0 \div \leq 4.0 \times 10^3$ cfu/m^3 and $<1.0 \times 10^0 \div 1.0 \times 10^3$ cfu/m^3, respectively. In case of indoor spaces where a high air quality is required (e.g. cleanrooms in action), not only the threshold limit values for HMA in the air are applied but recommended limits for microbial contamination of surfaces are used as well.

In 2004, the Expert Group on Biological Agents at the Polish Interdepartmental Commission for Maximum Admissible Concentrations and Intensities for Agents Harmful to Health in the Working Environment proposed the threshold limit values for the most common categories of microorganisms and bacterial endotoxin present in indoor air (Tab. 9) (Augustynska & Posniak 2007). These recommendations (which have been already applied in Poland for some years) can

Table 7. Admissible degree of microbial air contamination in utility rooms (Krzysztofik 1992).

Type of utility room	Admissible number of microorganisms in 1 m^3 if the air*		
	"Total" number of microorganisms on MPA medium	Number of hemolitic microorganisms on agar with blood	"Total" number of fungi on Sabouraud medium
Outdoor air (atmospheric)	3.0×10^3	1.0×10^2	1.0×10^3
Health service rooms:			
Operating room	1.0×10^2	0	0
Dressing room	1.5×10^2	0	5.0×10^1
Sick room	1.0×10^3	5.0×10^1	2.0×10^2
Rooms of residential homes:			
Kitchen and dining room	2.0×10^3	1.0×10^2	3.0×10^2
Saloon	1.5×10^3	5.0×10^1	2.0×10^2
Bedroom	1.0×10^3	5.0×10^1	1.0×10^2
School rooms:			
Lecture rooms	1.5×10^3	5.0×10^1	2.0×10^2
Exercise rooms	2.0×10^3	1.0×10^2	2.0×10^2
Gymnastic room	3.0×10^3	1.5×10^2	3.0×10^2
Productive rooms:			
Food industry	6.0×10^2	0	0
Meat industry	5.0×10^2	0	5.0×10^1
Fermentative industry	6.0×10^2	5.0×10^1	0
Pharmaceutical industry:			
a) dosing box	1.0×10^2	0	0
b) productive hall	3.0×10^3	5.0×10^1	1.0×10^2
Inventory rooms:			
Cowsheds	1.5×10^5	1.0×10^3	5.0×10^3
Rooms for calf	5.0×10^4	5.0×10^2	2.0×10^3
Stables	5.0×10^4	5.0×10^2	5.0×10^3
Piggeries	2.0×10^5	2.0×10^3	1.0×10^4
Rooms for piglets	5.0×10^4	1.0×10^3	5.0×10^3
Hen-houses	1.0×10^5	1.0×10^3	2.0×10^3

*concentrations determined by sedimentation method

be helpful not only in the assessment of exposure to HMA indoors but in the situations when the necessary prophylactic and/or preventive actions need to be undertaken.

Regarding outdoor air, in Poland two standards are still in force defining a degree of atmospheric air pollution with bacteria (the Polish Standard PN-Z-04111-02:1989 – "Air purity protection. Microbiological testing. Determination number of the bacteria in the atmospheric air (imission) with sampling by aspiration and sedimentation method") and fungi (the Polish Standard PN-Z-04111-03:1989 – "Air purity protection. Microbiological testing. Determination number of the fungi in the atmospheric air (imission) with sampling by aspiration and sedimentation method"). Unfortunately, both these regulations do not consider a modern (compatible with the current stage of knowledge) approach to the problem of microbial air quality, mistakenly allowing the use of a sedimentation method in quantitative assessment of bioaerosols. Prepared late in 2007 for the PKN the revision of these two standards is still "under discussion". In the methodical aspect, this revision unifies the proposals from three Polish Standards: PN-EN 13098, 14031 and 14042, resulting in a method of interpretation for the outdoor air quality assessment based on the empirically obtained results of HMA contamination. According to this alternation, the assessment of outdoor air contamination with bacteria, fungi and endotoxins can be performed based on both the quantitative and qualitative microbiological tests. The quantitative estimation assumes that the concentration of each bioaerosol compound should not exceed its threshold limit value proposed for the air of

Table 8. Hygienic standards for the air in ventilation and air-condition installations (a) and air-conditioned office rooms (b, c).

a) Italy (Cellai 1997):

Microorganisms	Concentration (cfu/m^3)	Hygienic assessment
Bacteria	pathogenic	occurrence inadmissible
	<50	very low concentration
	51–100	low concentration
	101–500	average concentration
	501–2000	high concentration
	>2000	very high concentration
Fungi	pathogenic	occurrence inadmissible, independently on the concentration
	>50	inadmissible when only one genus was present
	≤150	admissible when more than one genera were present
	≤300	admissible if the main genera were *Cladosporium* or *Philloplane*

b) Brazil (Neto and Siqueira 2000):

Microorganisms	Concentration (cfu/m^3)	Hygienic assessment
The sum of bacteria and fungi	$<7.5 \times 10^2$	Recommended level

c) United States, 2005 (Brandys and Brandys 2007):

Microorganisms	Concentration (cfu/m^3)	Hygienic assessment
Fungi	$<2.5 \times 10^2$	natural background level
	2.5×10^2–1.0×10^3	perhaps contamination source
	$>1.0 \times 10^3$	contamination source very possible

Table 9. Recommended threshold limit values for microorganisms and endotoxin in indoor air proposed by Expert Group for Biological Agents (Augustynska and Posniak 2007).

	Admissible concentration	
Microbial agent	Workplaces contaminated with organic dust	Residential and public utility premises
Mesophilic bacteria	1.0×10^5 cfu/m^3*	5.0×10^3 cfu/m^3
Gram-negative bacteria	2.0×10^4 cfu/m^3*	2.0×10^2 cfu/m^3
Thermophilic actinomycetes	2.0×10^4 cfu/m^3*	2.0×10^2 cfu/m^3
Fungi	5.0×10^5 cfu/m^3*	5.0×10^3 cfu/m^3
Agents from the risk groups 3 and 4	0 cfu/m^3	0 cfu/m^3
Bacterial endotoxin	200 ng/m^3 (2000 EU/m^3)*	5 ng/m^3 (50 EU/m^3)

*for the respirable fraction the proposed values should be lowered by a half; EU – Endotoxin Units

residential premises (as proposed by the Expert Group on Biological Agents – see Table 9), i.e. for the environment, in which most of the population spend most of their time during a day.

If the limit value for an individual bioaerosol compound is exceeded, the outdoor air should be recognized as "microbiologically polluted" and such conditions as "unacceptable". If the measured concentration is equal or lower than proposed value, the outdoor air quality should be recognized as "acceptable". The levels of microbial atmospheric air pollution proposed in this alternative standard (and shown in Table 10) have an arbitrary character, i.e. they were elaborated as a result of environmental measurements taking into consideration the potential harmfulness of each bioaerosol

Table 10. Proposals for the assessment of microbial contamination of outdoor air (Górny 2010).

	Level of microbial contamination of indoor air	
Bioaerosol compound	Acceptable	Unacceptable
Bacteria (the sum)	$\leq 5.0 \times 10^3$ cfu/m^3	$>5.0 \times 10^3$ cfu/m^3
Gram-negative bacteria	$\leq 2.0 \times 10^2$ cfu/m^3	$>2.0 \times 10^2$ cfu/m^3
Thermophilic actinomycetes	$\leq 2.0 \times 10^2$ cfu/m^3	$>2.0 \times 10^2$ cfu/m^3
Fungi	$\leq 5.0 \times 10^3$ cfu/m^3	$>5.0 \times 10^3$ cfu/m^3
Agents from the risk groups 3 and 4	0 cfu/m^3	0 cfu/m^3
Bacterial endotoxin	≤ 50 EU/m^3	>50 EU/m^3

EU – Endotoxin Units

constituent and they should be treated as facultative standards or auxiliary reference values (Górny 2010).

If the assessment of outdoor microbial air pollution was carried out with the simultaneous use of more than one method and the applied sampling techniques gave different concentrations of bioaerosols, then the final estimation of such assessment process should be based on the highest obtained concentration (presented as colony forming units or endotoxin units per 1 m^3 for microorganisms or endotoxins, respectively). If the bioaerosol sampling technique allows one to obtain the total number of microorganisms (i.e. both viable and non-viable together), such a method can be applied as a significant supplementary source of information about the levels of immunoreactive particulates in the air. In the case of microbial contaminants released into the air from special sources (e.g. sewage-treatment plants or landfills), an additional testing of indicatory microorganisms, specific for these sources, should be also performed. And vice versa, if the qualitative analysis of bioaerosols allows their detailed characteristics to the genus or species level, then the identification of indicatory microorganisms can give information about the point sources of microbial air pollutants (Górny 2010).

7 ALGORITHM OF ACTION IN EXPOSURE ASSESSMENT TO HARMFUL MICROBIAL AGENTS

The fundamental issue in exposure assessment to airborne HMA in indoor environment is compliance to the proper algorithm of action, which can be briefly described as follows: a) an understanding the nature of the observed phenomenon, b) a correct selection of the correct measurement method and its proper execution, c) a determination of indicators of HMA exposure, and d) a comparison of the obtained measurement results with the proposed hygienic standards/threshold limit values. The results of such exposure assessment would be a solid base to perform a proper risk assessment (including occupational risk assessment, if indoor measurements were carried out in a working environment) and to select the suitable protective equipment.

REFERENCES

ACGIH – American Conference of Governmental Industrial Hygienists. 1989. *Guidelines for Assessment of Bioaerosols in Indoor Environment*. Cincinnati: ACGIH.

ACGIH – American Conference of Governmental Industrial Hygienists. 1995. Macher, J.M., Chatigny, M.A., Burge, H.A. Sampling airborne microorganisms and aeroallergens. In B.S. Cohen & S.V. Hering (eds), *Air Sampling Instruments for Evaluation of Atmospheric Contaminants*: 589-617. Cincinnati: ACGIH.

AFNOR – Association Francaise de Normalisation. 2003. NF S 90-351 – Etablissement de Santé: Salles Propres et Environnements Maîtrisés et Apparentés. Paris: AFNOR (in French).

AIHA – American Industrial Hygiene Association. 1986. *Biosafety Committee: Biohazards Reference Manual*. Washington DC: AIHA.

AIHA – American Industrial Hygiene Association. 1989. The practitioner's approach to IAQ investigations. In D.M. Weekes & R.B. Gammage (eds), *Proc. Indoor Air Quality Intern. Symp.*: 43 & 66. Fairfax: AIHA.

American Academy of Orthopedic Surgeons. 1976. In Charkowska, A. 2008. Ensuring cleanliness in operating theatres. *JOSE* 14: 447–453.
American Pharmacopoeia. 1992. *USP, chapter 1116: US Federal Standard 209E: Airborne Particulate Cleanliness Classes for Cleanrooms and Clean Zones.*
Augustynska, D. & Posniak, M. (eds). 2007. *Harmful Agents in the Working Environment – Admissible Values.* Warsaw: CIOP-PIB (in Polish).
Berk, J.B., Boyan, T.A., Brown, S.R., Ko, I., Koonce, J.F., Loo, B.W., Pepper, J.H., Robb, A.W., Strong, P.C., Turiel, I. & Young, R.A. 1980. Field monitoring of indoor air quality. In 1979 *Annual Report of the Energy and Environment Division.* Berkeley: Lawrence Berkeley Laboratory, University of California.
Binnie, P.W.H. 1990. Biological pollutants in the indoor environment. In J.G. Kay, G.E. Keller, J.F. Miller (eds): *Indoor Air Pollution.* Chelsea: Lewis Publishers.
Bourdillon, R.B., Lidwell, O.M. & Thomas, J.E. 1941. A slit sampler for collecting and counting the airborne bacteria. *Journal of Hygiene* 41: 197–224.
Bowser Technical Inc. 1991. *Testing of Older Houses for Microbiological Pollutants.* Ottawa: Canada Mortgage and Housing Corporation.
Brandys, R.C. & Brandys, G.M. 2007. *Worldwide Exposure Standards for Mold and Bacteria.* 7th Edition. Hinsdale: OEHCS, Inc., Publications Division.
Bujwid, O. 1894. *Bakteryje w Powietrzu: Sposoby Badania, Znaczenie i Opis Pospolicie Znajdowanych Bakterii.* Warsaw: Towarzystwo Lekarskie Warszawskie 1894 (in Polish).
Burge, H.A. 1990. Bioaerosols: prevalence and health effects in the indoor environment. *Journal of Allergy and Clinical Immunology* 86: 687–701.
Buyanov, V.V., Sergeyuk, N.P. & Kaplynov, Y.V. 1990. Gigenicheskoye normirovanye soderzhanya v vozdukhe uslovno-patogennykh mikroorganizmov. *Gigiena truda i professional'nye zabolevaniia* 8: 27–30 (in Russian).
Castellan, R.M., Olenchock, S.A., Kinsley, K.B. & Hankinson, J.L. 1987. Inhaled endotoxin and decreased spirometric values. An exposure-response relation for cotton dust. *New England Journal of Medicine* 317: 605–610.
CEC – Commission of the European Communities. 1993. *Indoor Air Quality & its Impact on Man: Report No. 12: Biological Particles in Indoor Environments.* Brussels-Luxembourg: ECSC-EEC-EAEC.
Cellai, G. 1997. La protettazione delle condotte dell'aria ai fini della manutenzione. Documenti e riferimenti normativi in relazione all'IAQ, Seconda parte. *AICRR* 8, 789–795 (in Italian).
China: Ministry of Health. 1966 & 2001 (www2u.biglobe.ne.jp).
Clark, C.S. 1985. Report on prevention and control. Proc. of an International Workshop Health Effects of Organic Dusts in the Farm Environment held in Skokloster, Sweden, April 23–25, 1985. *American Journal of Industrial Medicine* 10: 267–273.
Clark, G. 2005. Proposed air and source sampling numerical guidelines for fungal remediation. In R.C. Brandys & G.M. Brandys, *Worldwide Exposure Standards for Mold and Bacteria. 7th Edition,* 2007. Hinsdale: OEHCS, Inc., Publications Division.
CMHC – Canada Mortgage and Housing Corporation, Paracel Laboratories. 1988. *Determination of Fungal Propagules in Indoor Air.* Ottawa: CMHC.
Cutter Information Corp.: Etkin, D.S. 1994. *Indoor Air Quality Update: Biocontaminants in Indoor Environments.* Arlington: CIC.
DECOS – Dutch Expert Committee on Occupational Standards. 1989. *Research Methods in Biological Indoor Air Pollution.* Den Haag: DECOS.
DECOS – Dutch Expert Committee on Occupational Standards. 1998. *Endotoxins: Health Based Recommended Occupational Exposure Limit.* Den Haag: DECOS.
Directive 2000/54/EC of the European Parliament and of the Council of 18 September 2000 on the Protection of Workers from Risks Related to Exposure to Biological Agents at Work. *Official Journal of European Communities L.* 262/21, Brussels.
Donham, K. 1988. In S. DeBoer & W.D. Morrison, *The Effects of the Quality of Environment in Livestock Buildings on the Productivity of Swine and Safety of Humans.* A Literature Review, Guelph: University of Guelph, Canada.
Donham, K., Cumro, D., Reynolds, S.J. & Merchant, J.A. 2000. Dose-response relationships between occupational aerosol exposures and cross-shift declines of lung function in poultry workers: recommendations for exposure limits. *Journal of Occupational and Environmental Medicine* 42: 260–269.
Donham, K.J. 1991. Association of environmental air contaminants with disease and productivity in swine. *American Journal of Veterinary Research* 52: 1723–1730.
Douwes, J., Thorne, P., Pearce, N., Heederik, D. 2003. Bioaerosol health effects and exposure assessment: progress and prospects. *Annals of Occupational Hygiene* 47: 187–200.

Dutkiewicz, J. & Molocznik, A. 1993. *Zweryfikowana dokumentacja NDS dla pyłów pochodzenia roslinnego i zwierzecego*. Lublin: Instytut Medycyny Wsi (in Polish).
Erman, M.I., Eglite, M.E., Olefir, A.I., Kalinina, L.N. 1989. Aerogennaya mikroflora zhivotnovodczeskikh i pticevodczeskikh proizvodstvennykh pomeshchennyi, kriterii eyo vrednogo deistva i gigyenicheskaya reglamentatsia", *Gigiena truda i professional'nye zabolevaniia* 4: 19–22 (in Russian).
European Commission Guide to Good Manufacturing Practice (EC GMP). 2003. *Manufacturing of Sterile Medicinal Products, Revision to Annex 1*. Brussels: EC.
Finnish Institute of Occupational Health. 1997. *Sosiaali- ja Terveysministeriön Oppaassa, Sisäilmaohjeessa on Sisäilman Mikrobeille Esitetty Seuraavia Ohjearvoja* (in Finnish).
Germany: standard DIN 1946-4. 1999 (replaced in 2008). Ventilation and Air Conditioning - Part 4: Ventilation in Buildings and Rooms of Health Care, Berlin: DIN.
Godish, T. 1991. *Indoor Air Pollution Control*. Chelsea: Lewis Publishers.
Górny, R.L. & Dutkiewicz, J. 2002. Bacterial and fungal aerosols in indoor environment in Central and Eastern European countries. *Annals of Agricultural and Environmental Medicine* 9: 17–23.
Górny, R.L. 2004. Biological harmful agents: standards, recommendations and proposal of admissible values. *PiMOSP* 3: 17–39 (in Polish).
Górny, R.L. 2010. Biological aerosols – the role of hygienic standards in the environment and health protection. *Environmental Medicine* 13: 41–51 (in Polish).
Haglind, P. & Rylander, R. 1984. Exposure to cotton dust in an experimental cardroom. *British Journal of Industrial Medicine* 41: 340.
Heida, H., Bartman, F., Van der Zee, S. 1995. Occupational exposure and indoor air quality monitoring in a composting facility. *American Industrial Hygiene Association Journal* 56: 39–43.
Holmberg, K. 1991. Mould growth inside buildings. In B. Flannigan, E.M. McCabe, F. McGarry (eds), Allergenic and toxigenic micro-organisms in houses. *Journal of Applied Bacteriology, Symposium Supplement* 70: 61S–73S.
Indoor Air Quality Association Inc. 1995. *IAQ Standard No. 95-1 Recommended for Florida*. Longwood, IAQ Association Inc.
Institut für Baubiologie & Oekologie. 2003. Baubiologie Maes/IBN – Standard of building biology testing methods and guidelines SBM-2003. Neubeuern: IBN.
International Committee of Contamination Control Societies – ICCCS. 1991. In W. Whyte, *Cleanroom Design*. Baffins Lane: John Wiley and Sons Ltd.
Italian Institute of Health. 1995. In F. Rabino, Strategie di sterilizzazione e sanitizzazione degli impianti tecnici dei locali nell'ambito ospedaliero. *Convegno AICARR. La gestione dell'impiantistica ospedaliera: strategia, aspetti tecnici ed energetici, 22.06.1995, Abano Terme, Italy*, 17–56 (in Italian).
Kalliokoski, P. 2003. Risks caused by airborne microbes in hospitals – source control is important. *Indoor and Built Environment* 12: 41–46.
Klanova, K. & Drahonovska, H. 1999. The concentrations of mixed populations of fungi in indoor air: rooms with and without mould problems; rooms with and without health complaints. *Proc. Indoor Air'99*, 1: 920–924. London: Construction Research Communications Ltd.
Kruczkowski, P. 1984. *Wytyczne projektowania szpitali ogólnych. Instalacje sanitarne. Z. 5. Wentylacja i klimatyzacja*. Warsaw: Biuro Studiów i Projektów Sluzby Zdrowia (in Polish).
Krzysztofik, B. 1992. *Microbiology of the air*. Warsaw: Wydawnictwa Politechniki Warszawskiej (in Polish).
Lacey, J. & Crook, B. 1988. Fungal and actinomycetes spores as pollutants of the workplace and occupational allergens. *Annals of Occupational Hygiene* 32: 515–533.
Laitinen, S. 1999. *Exposure to Airborne Bacteria in Occupational Environments*. Ph.D. thesis. Kuopio: University of Kuopio.
Madsen, A.M., Mårtensson, L., Schneider, T. & Larsson, L. 2004. Microbial dustiness and particle release of different biofuels. *Annals of Occupational Hygiene* 48: 327–338
Malmberg, P. 1991. Microorganisms. In B. Beije & P. Lundberg (eds): *Criteria Documents from the Expert Group*, 39–69. Solna: Arbets Milio Institutet.
Malmros, P., Sigsgaard, T. & Bach, B. 1992. Occupational health problems due to garbage sorting. *Waste Management Research* 10: 227–234.
Martinez, K.F., Rao, C.Y. & Burton, N.C. 2004. Exposure assessment and analysis for biological agents. *Grana* 43: 193–208.
Miller, J.D., Laflamme, A.M., Sobol, Y., Lafontaine, P. & Greenhalgh, R. 1988. Fungi and fungal products in some Canadian houses. *International Biodeterioration* 24: 103–120.
Morey, P.R., Hodgson, M.J., Sorenson, W.G., Kullman, G.J., Rhodes, W.W. &Visvesvara, G.S. 1984. Environmental studies in moldy office buildings: biological agents, sources and preventative measures. *Annals of ACGIH* 10: 21–35.

Neto, F.R.A. & Siqueira, L.F.G. 2000. Guidelines for indoor air quality in offices in Brazil. *Proc. Healthy Buildings* 4: 549–554.
Nevalainen, A. 1989. *Bacterial Aerosols in Indoor Air*. Ph.D. thesis. Helsinki: National Public Health Institute.
New York City Guidelines on Assessment and Remediation of Fungi in Indoor Environments, New York, Department of Health. 1994. In E. Johanning & C.S. Yang (eds), *Fungi and bacteria in indoor air environment, 201–207. Proc. intern. conf., 6–7 October 1994, Saratoga Spring*, New York.
Nordic Council: Criteria Documents from the Export Group. 1993. In T. Nathanson*: Indoor Air Quality in Office Buildings: A Technical Guide*. Ottawa: Department of National Health and Welfare.
Ohgke, H., Geers, A. & Beckert, J. 1987. Fungal load of indoor air in historical and newly constructed buildings used by public services. *Proc. Indoor Air* 1: 681–684.
Ordinance 716 (22.04.2005, with its alteration: Ordinance 288 (29.02.2008)) of the Ministry of Health on Harmful Biological Agents in Working Environment and Health Protection of Workers Occupationally Exposed to These Agents. Warsaw (in Polish).
Ordinance 1882 (03.12.2002) of the Ministry of Health on Requirements of Good Manufacturing Practice. Warsaw (in Polish).
Palchak, R.B., Cohen, R., Ainslie, M. & Hoerner, C.L. 1988. Airborne endotoxin associated with industrial-scale production of protein products in gram-negative bacteria. *American Industrial Hygiene Association Journal* 49: 420–421.
Rabino, F. 1995. Strategie di sterilizzazione e sanitizzazione degli impianti tecnici dei locali nell'ambito ospedaliero. *Convegno AICARR. La gestione dell'impiantistica ospedaliera: strategia, aspetti tecnici ed energetici, 22.06.1995, Abano Terme, Italy*: 17–56 (in Italian).
Reponen, T., Nevalainen, A., Jantunen, M., Pellikka, M. & Kalliokoski, P. 1990. Proposal for an upper limit of the normal range of indoor air bacterial and fungal spores in subarctic climate, *Proc. Indoor Air* 2: 47–50. Ottawa: Canada Mortgage and Housing Corporation.
Reynolds, S.J., Streifel, A.J., McJilton, C.E. 1990. Elevated airborne concentrations of fungi in residential and office environments. *American Industrial Hygiene Association Journal* 51: 601–604.
Russian Federation: State Committee for Hygiene and Epidemiological Surveillance. 1993. Maximum Allowable Concentrations of Harmful Substances. Moscow: SCHES.
Rylander, R. 1987. The role of endotoxin for reactions after exposure to cotton dust. *American Journal of Industrial Medicine* 12: 687–697.
Sartorius. 1995. *Wichtige Aspekte zur Luftkeimsammlung in Reinräumen mit der Gelatine-Membranfilter-*Methode (in German).
Schweizerische Krankenhausinstitut (SKI). 1987. *Richtlinie für den Bau, Betrieb und die Überwachung von lüftungstechnischen Anlagen in Spitälern*. Aarau: Krankenhausinstitut (in German).
Smid, T., Heederik, D., Mensink, G., Houba, R. & Boleij, J.S. 1992. Exposure to dust, endotoxins, and fungi in the animal feed industry. *American Industrial Hygiene Association Journal* 53: 362–368.
Solomon, W.R. & Burge, H.P. In P.J. Walsh, E.D. Dudney, E.D. Copenhaver (eds), *Indoor Air Quality*, 174–191. Boca Raton: CRC Press.
Standard NASA – NHB 5340.2. 1993. In W. Whyte & S. Bresin, *Les Salles Proper. Maitriser la Contamination: Pourquoi? Comment?* Ivry-sur-Seine: PYC Edition.
Suva – Caisse nationale suisse d'assurance en cas d'accidents. 2005. *Valeurs Limites d'Exposition aux Postes de Travail 2005. Référence: 1903.f. February 2005*. Suva Switzerland (in French).
Topley, W. & Wilson, G. 1955. *Principles of Bacteriology and Immunology*: 2275.
US OSHA – United States Occupational Safety and Health Administration. 1992. *Technical Manual*. Washington DC: OSHA.
Wells, W.F. 1995. Airborne Contagion and Air Hygiene. Cambridge: Harvard University Press.
WHO – World Health Organization. 1988. *WHO Regional Publications, European Series No. 31: Indoor air quality: biological contaminants. Report on a WHO meeting, Rautavaara, 29 August–2 September 1988*. Copenhagen: WHO Regional Office for Europe.
WHO – World Health Organization. 1998. *WHO Regional Publications, European Health for All Series No. 5: HEALTH21: an introduction to the health for all policy framework for the WHO European Region*. Copenhagen: WHO Regional Office for Europe.
WHO – World Health Organization. 2000. *The Right to Healthy Indoor Air. Report on a WHO meeting, Bilthoven, 15–17 May 2000*. Copenhagen: WHO Regional Office for Europe.
Whyte, W. & Coll, A. 1995. In F. Rabino, Strategie di sterilizzazione e sanitizzazione degli impianti tecnici dei locali nell'ambito ospedaliero. *Convegno AICARR. La gestione dell'impiantistica ospedaliera: strategia, aspetti tecnici ed energetici, 22.06.1995, Abano Terme, Italy*: 17–56 (in Italian).
Yang, C.S., Hung, L.L., Lewis, F.A. & Zampiello, F.A. 1993. Airborne fungal populations in non-residential buildings in the United States. *Proc. Indoor Air* 4: 219–224.

Management of Indoor Air Quality – Dudzińska (ed)

Seasonal variation and size distribution of bioaerosols in an air-conditioned auditorium – a case study

A. Rożej, M.R. Dudzińska & U. Gąska-Jędruch
Faculty of Environmental Engineering, Lublin University of Technology, Lublin, Poland

ABSTRACT: HVAC systems may be responsible for the secondary pollution of indoor environments by bioaerosols. Therefore, the changes in microbiological air quality were determined in the mechanical air supply system of an auditorium at the Faculty of Environmental Engineering, Lublin University of Technology, Poland.

Air samples were collected using a six stage Andersen impactor with agar plates at four sites: outdoor next to the fresh air inlet, supply duct, the central zone of the vacant auditorium and the exhaust duct for heterotrophic bacteria, staphylococci and fungi count.

Periodical switch off of the air conditioning caused an increase of humidity and the mould content in the supply duct; its activation reduced the air quality in the auditorium and air supply system.

The bacteria content in auditorium air was more the result of human activity than outdoor air quality.

Keywords: airborne bacteria, bioaerosols, fungi, indoor air, HVAC

1 INTRODUCTION

Bioaerosols play a significant role in indoor air pollution and have a direct impact on human health associated with respiratory disease, asthma (Ross et al. 2000), sick building syndrome (Crook & Burton 2010, Shoemaker & House 2005) and reduced productivity (Crook & Sherwood-Higham 1997).

Directive 2000/54/EC of the European Parliament and of the Council of 18 September 2000 on the protection of workers from risks related to exposure to biological agents at work likewise European Standards (EN 13098:2000, EN 14031:2003) defined the biological agents and methods of their measurement in work places but did not give any required limits for microbial cell concentrations (Albrecht et al. 2007). However, oddly, neither recommended threshold values in indoor air nor permissible exposure limits for occupation safety have been recommended in EU documents. A similar situation occurs in individual countries such as Poland. Although in some countries there are regulations which govern the microbial contamination levels of indoor air they apply only to special rooms such as surgical operating theatres (NF-S-90-351 1987; NHS Estates 1994).

Górny (2004) specified the range of limits of microbial concentrations proposed by numerous organisations, institutions, groups of specialists and individual researchers in various parts of the world during the last hundred years:

a) For a total number of bacteria $\leq 1.0 \cdot 10^3 \div \leq 7.0 \cdot 10^3$ CFU/m^3 in dwellings and non-industrial work places, and $\leq 7.5 \cdot 10^2 \div 1.0 \cdot 10^7$ CFU/m^3 in industrial work places. There are no safe concentrations for pathogenic organisms and the limit for their concentration should be estimated as 0 CFU/m^3.
b) For Gram-negative bacteria $1.0 \cdot 10^3 \div 2.0 \cdot 10^4$ CFU/m^3 in industrial work places.
c) For bacterial endotoxins $0.005 \div 0.2$ μg/m^3 in industrial work places.
d) For fungi $1.0 \cdot 10^1 \div 1.0 \cdot 10^4$ CFU/m^3 in dwellings and non-industrial work places, and $< 1.0 \cdot 10^2 \div 1.0 \cdot 10^7$ CFU/m^3 in industrial work places.
e) For subtilisine 0.06 μg/m^3 in industrial work places.

f) For house dust allergens:
- house dust mites *Der p* I: 2 ÷ < 15 μg/g,
- cat *Fel d* I: < $1.0 \cdot 10^4$ ng/g,
- dog *Can f* I: < $1.0 \cdot 10^5$ ng/g.

According to data in the published literature (Nevalainen & Seuri 2005), the indoor concentrations of viable fungi under 100 CFU/m^3 are considered 'low' and those over 1000 CFU/m^3 are regarded as 'high'. The outdoor levels of airborne microflora, and specially fungi reveals seasonal changes, with a peak in summer (Liao & Luo 2005, Medrela-Kuder 2003, Parat et al. 1997).

The epidemiological impact of airborne microorganisms and particles was documented as size dependent (Liao & Luo 2005, Sattar & Ijaz 1987). The large particles are trapped in the upper respiratory tract (nose and nasopharynx) while particles <6 μm in diameter are transported to the lung (Zeterberg 1973 quoted after Stetzenbach 1998). While Kim & Kim (2007) as well as Wang et al. (2010) calculated the "respirable fraction" of bacteria and fungi as the sum of their concentrations in fractions of particles smaller than 4.7 μm in diameter (stages 3–6 of six-stage Andersen impactor). The greatest retention in the lung alveoli was reported for particles 1–2 μm in diameter (Sattar & Ijaz 1987) or smaller (Liao & Luo 2005).

The aerodynamic sizes of spores are typical for specific microorganism species but may vary due to the physical conditions of the indoor environment such as air velocity and humidity (Pasanen et al. 1991; Reponen et al. 1996, 1998). An increase of the contribution of small size particles in air samples was observed in summer (Wang et al. 2010) when desiccation of the bioaerosols can proceed faster.

Tham & Zuraimi (2005) as well as earlier Palmgren et al. (1986) showed that the percentages of particles containing viable bacteria at different sizes were all very low (<1%). Similarly culturable fungi represent only about 1% of the fungal particles present in the air (Toivola et al. 2002).

Heating, Ventilation and Air-Conditioning (HVAC) may be responsible for the production and dispersion of microorganisms in buildings (Parat et al. 1997). Low efficiency filters, cold mist humidifiers with water recirculation, water condensation areas and large recirculation of air can lead to proliferation of microorganisms and secondary air pollution (Parat et al. 1997).

The aim of the study was to determine the changes in microbiological air quality in the mechanical air supply system of a university auditorium under various conditions of outdoor air parameters, on/off status of the air-conditioning system and the presence of occupants.

2 MATERIALS AND METHODS

Measurements of indoor bioaerosol concentrations were performed in the air-conditioned auditorium at the Faculty of Environmental Engineering of the Lublin University of Technology. The auditorium (area 300 m^2, air-space 1200 m^3) has been in use since October 1, 2007.

The HVAC system regulated the temperature and humidity of air in the lecture hall as well as removing aerosols by fabric bag filters EU4. Ac system filters were changed every three months. For these studies the set of filters was operated since July 2008 till January 2010, when it was replaced by new one. The air velocity in the air supply duct was 5 m/s. The system supplied 7000 m^3/h of air and it provided an air exchange level of ca. 6 h^{-1}. The exhaust air was partially recirculated (recirculation rate 50%) and mixed with the fresh air.

Examinations were conducted in eight measurement series in the period between 15th of July 2009 and 10th of March 2010. Five measurements were performed during the summer interval in the academic cycle to mitigate the effect of humans on air quality. The succeeding three series took place during the academic year, but the air sampling was carried out in the empty auditorium.

Details of the conditions of every experimental series are gathered in Table 1.

Air samples were collected by using a six stage Andersen impactor (range of particle size from 0.65 to >7 μm) with agar plates at four sites: outdoor next to fresh air inlet, supply duct behind the air conditioning station and steam humidifier, the cental zone of the unoccupied auditorium and exhaust duct. Outdoor air sampling was impossible in January and March 2010 because of very low outdoor temperature (<0°C).

The sampler operated at 28.3 dm^3/min for 1–10 min, depending on the expected cell concentration at the respective sampling site. Between measurements, the sampler was thoroughly cleaned

Table 1. The detailed conditions of successive measurements.

Series	Time	Measurement Conditions
I	15.07.2009	The measurement immediately after the end of the academic term; air conditioning off
II	22.07.2009	One week after the end of the academic session, air conditioning off
III	29.07.2009	After one week of air conditioning
IV	01.09.2009	Immediately after activation of air conditioning after four weeks holidays
V	23.09.2009	After three weeks of air conditioning operation
VI	26.11.2009	After three months of air conditioning operation
VII	29.01.2010	After replacement of air filters
VIII	10.03.2010	After two months of air conditioning with new air filters

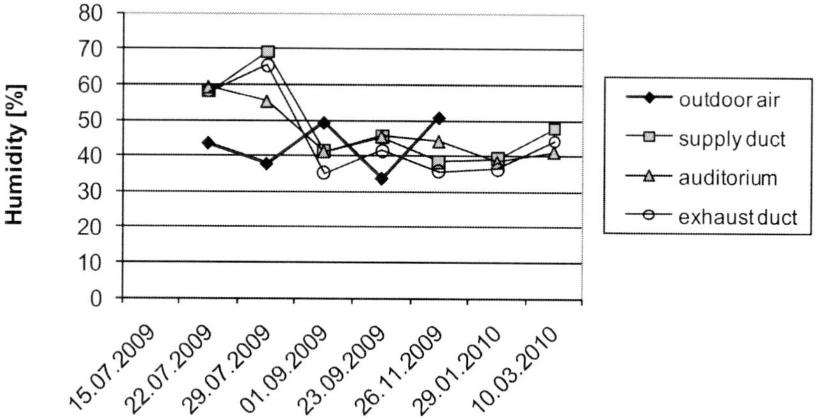

Figure 1. The relative humidity of air in the air supply system of the auditorium.

by swabbing with 70% ethanol. At the auditorium air samples were collected at the sampling height of 1.2 m, near the human breathing zone.

At each sampling site air samples were collected on four sets of Petri dishes with appropriate media for heterotrophic bacteria, staphylococci and fungi. The following media were used for colony counting:

- Tripticase Soy Agar (TSA) for detection of heterotrophic bacteria
- Baird-Parker Agar for detection of staphylococci
- Sabourand Agar with chloramphenicol for detection of fungi
- Chapek's Agar for detection of fungi.

The Petri dishes with TSA and B-P agar were incubated in 37°C for 24–48 hours and the plates with Chapek's and Sabourand media were incubated in 28°C for 7–14 days. The results of measurement are presented as the number of Colony Forming Units (CFU) per cubic meter of air.

The bacterial and fungal colonies were identified morphologically and biochemically (API Tests – Biomerieux for bacteria).

Simultaneously with air sampling the temperature and the relative humidity were measured by using a Datalogger Lab-L LB520 exact to 0.1°C and 2% of the relative humidity.

3 RESULTS AND DISCUSSION

The physical conditions of air in the auditorium and its air supply system were stable during the experiment. Temperature did not vary more than 2–3°C between 22.1 and 25.3°C and the relative humidity ranged from 38.4–45.5% in the lecture hall (Figs 1–2). Only when the air conditioning was out of action in July 2009 the humidity in air supply system rose to 69 (series II).

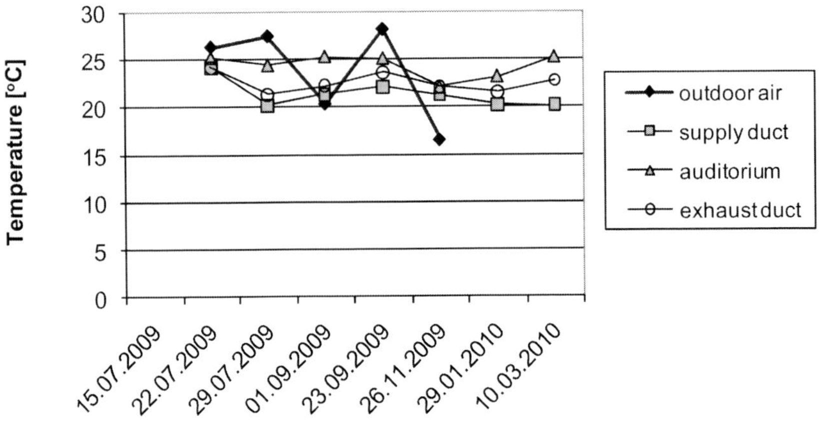

Figure 2. The temperature of air in the air supply system of the auditorium.

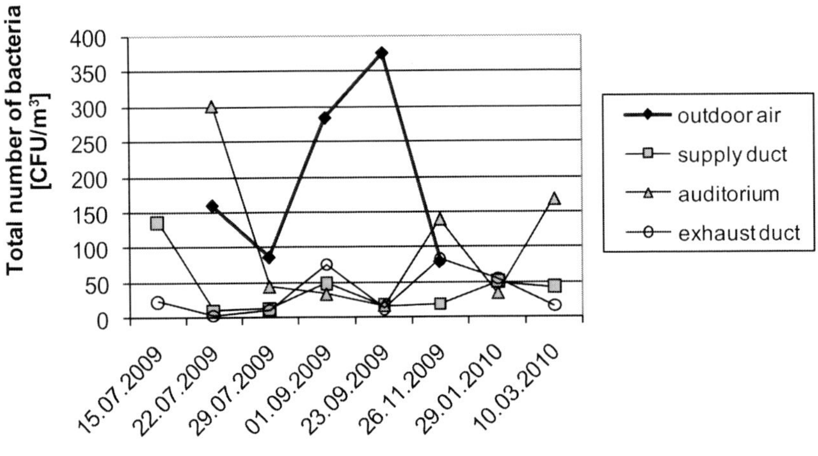

Figure 3. Number of heterotrophic bacteria in the air supply system of the auditorium (the sum of colonies obtained from six impactor stages).

The total number of heterotrophic bacteria in the air supply system of the lecture hall remained at a low level while the outdoor air contained ten-fold more bacteria during summer than indoor air (Fig. 3). It did not exceed 50 CFU/m^3 in the supply- and 82 CFU/m^3 in exhaust duct. More bacteria were observed in the auditorium air sampled during the academic year (November 2009 and March 2010). The highest number of bacteria in the auditorium air was recorded in 22 July 2009, when the air conditioning was out of action and the air in supply ducts contained only a few CFU per cubic meter.

Most of those bacteria were grampositive cocci (Fig. 4) and 85% of the counted colonies were recorded in fraction 3.3–4.7 μm of bioaerosols (Fig. 5). Because of an excess of colonies and consolidated growth on plates with TSA medium it was impossible to count the total number of heterotrophic bacteria and their individual contribution to various fractions of bioaerosols. The total number of bacteria was assumed to be 300 CFU/m^3, not less than the number of grampositive cocci (Fig. 3), but the size distribution of particles on 22 July 2009 is not shown in Figure 5. Presumably students may have occupied the auditorium prior to sampling. In other measurements the grampositive cocci were observed in only a few particles per cubic meter.

Most of the bacterial isolates taken from the auditorium and supply ducts air belonged to genera *Micrococcus* and *Bacillus* but *Staphylococcus* and *Actinomyces* were also present. *Micrococcus luteus* was observed most frequently in the air supply system. The following other grampositive cocci were identified in air samples: *Staphylococcus aureus*, *S. hominis* and *S. haemolyticus*. Among

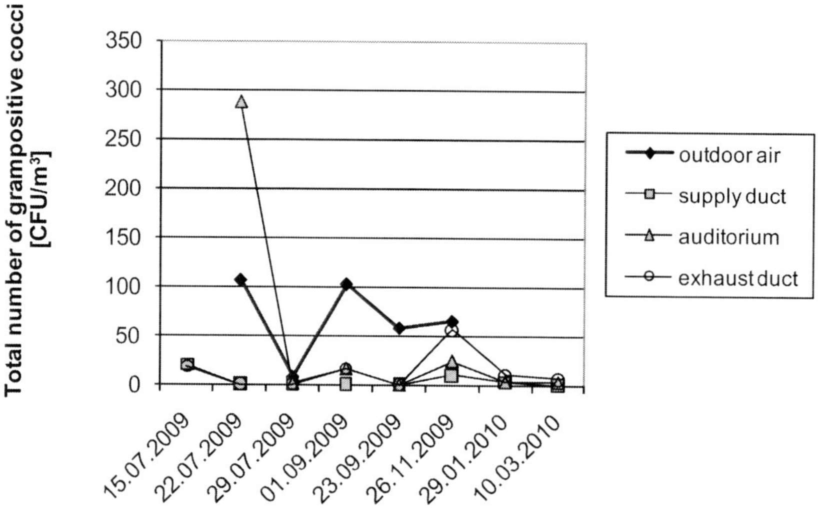

Figure 4. Number of grampositive cocci in the air supply system of the auditorium (the sum of colonies obtained from six impactor stages).

grampositive bacilli in most cases *Bacillus pumilis* was observed, less frequently *B. megaterium* and *B. licheniformis*.

During summer 2009 the outdoor air contained a large amount of fungal spores, but its number fluctuated depending on weather conditions e.g. temperature, humidity, atmospheric precipitation. The highest number of mould in outdoor air was observed during hot, sunny days (Fig. 6). The content of fungal spores in the air supply system of auditorium did not vary so evidently. The supply duct air contained 7.5-fold less fungal spores than outdoor air on 22 July and 5.6-fold less on 23 September 2009. In the autumn both the outdoor and indoor air contained fewer molds.

The most disconcerting result of air sampling was observed 29 July 2009 when a huge amount of mold was recorded in the supply duct air (1919 CFU/m^3). It was at the time when the air conditioning was re-started after two weeks shut-down. During that period the humidity had increased in the air supply system (Fig. 1) facilitating mold growth. On re-starting, this massive airflow may have caused the large scale dispersal of fungal spores in the duct. Interestingly, the mold content in the auditorium air did not rise.

The replacement of spent filters by new ones in January 2010 decreased the number of fungal spores to 3 CFU/m^3 in the supply air. Three months later the quality of air in both the lecture hall and the supply system were as good as at the moment followed the filter exchange.

Among fungi isolated from both outdoor and indoor air the following genera dominated: *Cladosporium*, *Aspergillus* and *Penicillium*. The genus *Aspergillus* was represented most frequently by *A. niger* and *A. fumigatus*, but other species such as *A. flavus*, *A. versicolor* and *A. sclerotiorum* were also observed. In all air samples other fungal genera were identified: *Fusarium*, *Mucor*, *Absidia*, *Geotrichum*, *Alternaria*, *Chaetomium* and *Rhisopus*.

Outdoor air contained fungal spores in all sizes of particles distinguished using the Andersen impactor, but the highest levels of spores were in the size range of 2.1–4.7 μm (Fig. 7). Accordingly these air samples made up the highest contribution of particles larger than 7 μm (7–17%). The indoor air both in air supply ducts and in the lecture hall contained fractions 1.1–3.3 μm primarily and finally the smallest particles in the range 0.65–2.1 μm dominated in air sampled in September and November 2009. In the critical measurements of 22 July 2009 when a huge number of fungal spores was transported in air through supply duct more than 99% of particles were smaller than 7 μm in diameter. The size distribution of airborne bacteria and fungi observed in the present studies confirmed literature data such as that of Meklin et al. (2002) who observed the highest fungal levels in the size range of 1.1–4.7 μm in Finnish school buildings constructed of two materials (wood and concrete). Wang et al. (2010) also recorded that the highest fraction for bacteria and fungi

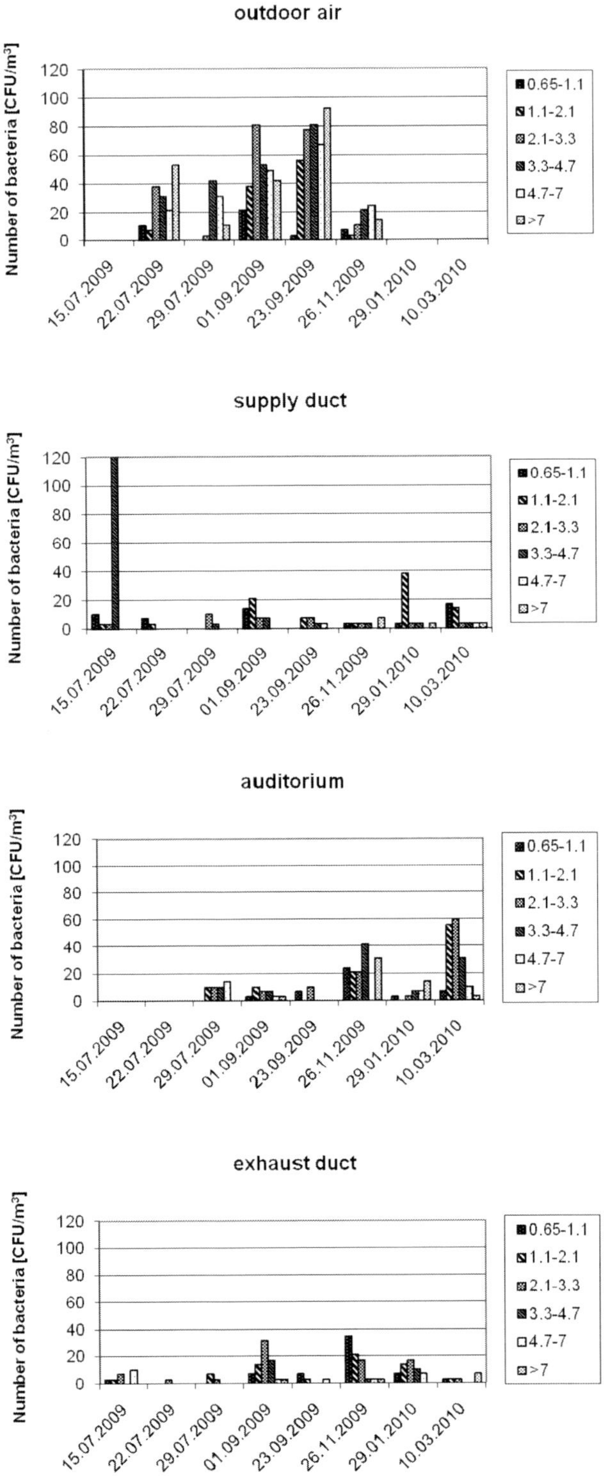

Figure 5. Particle size (μm) distributions of heterotrophic bacteria in the air supply system of the auditorium.

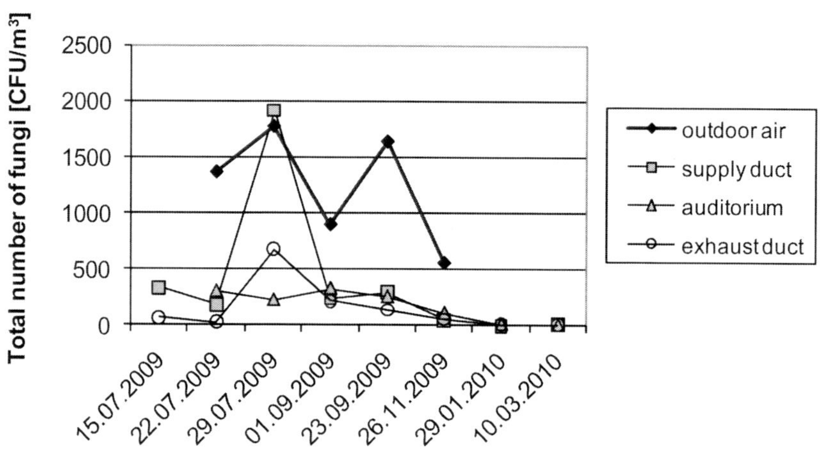

Figure 6. The number of viable fungi in the air supply system of the auditorium (the mean value of the sum of colonies obtained from six impactor stages on two different media).

occurring in commuting trains air was in particles 1.1–2.1 μm in diameter and its contribution in bioaerosols increased during summer.

This is a highly undesirable situation increasing human infection risk because such small particles pass through the upper respiratory tract and can be deposited in the trachea, bronchi and bronchioles with the greatest retention in the lung alveoli as reported by Sattar & Ijaz (1987) and Liao & Luo (2005).

The outdoor levels of airborne fungi revealed seasonal changes, peaking in the summer, as typically described in the literature (Liao & Luo 2005, Medrela-Kuder 2003, Parat et al. 1997). Because of negative outside temperatures it was impossible to sample outdoor air and count bioaerosols in January and March 2010.

The predominant indoor and outdoor genera of fungal spores were *Aspergillus*, *Penicillium* and *Cladosporium* as reported in several studies of the air in urban homes (Górny & Dutkiewicz 2002, Lee et al. 2006), schools as well as an academic lecture hall (Medrela-Kuder 2003). Bacterial isolates from indoor air most frequently belonged to genera: *Micrococcus*, *Bacillus* and *Staphylococcus* likewise in agreement with the literature data (Górny & Dutkiewicz 2002). The bacteria content in auditorium air was the result rather of human activity (autumn and winter measurements) than outdoor air quality.

4 CONCLUSIONS

- The outdoor air bioaerosols revealed seasonal variations. In agreement with the literature, an elevated number of bacteria and fungi were recorded in summer.
- The total number of heterotrophic bacteria and fungi in the air supply system of the lecture hall remained at a low level while the outdoor air contained ten-fold more microorganisms during summer than the indoor air.
- Outdoor air contained bioaerosols in all size ranges as distinguished using an Andersen impactor. The highest levels of fungi were in the particle size range of 2.1–4.7 μm while the indoor air contained fungi mainly in 1.1–3.3 μm fraction.
- Bacteria were recorded in a wide range of particle sizes in the indoor air. The higher contribution of larger fractions was observed during occupancy of the auditorium.
- Most of bioaerosols in the indoor air containing bacteria and fungi were respirable particles below 7 μm in diameter. Particles >7 μm were more frequent in outdoor air – up to 17% of the total for fungi and 33% for bacteria.
- The content of bacteria in the auditorium air was the rather the result of human activity than outdoor air quality.

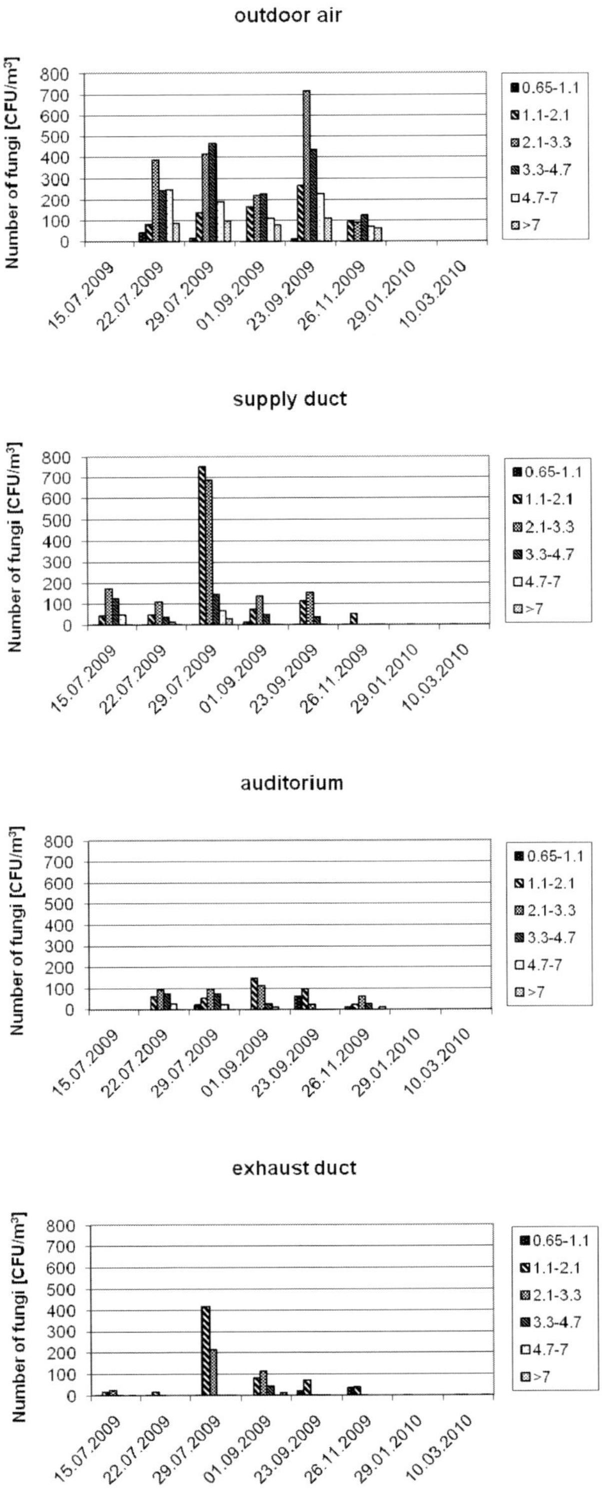

Figure 7. Particle size (μm) distributions of viable fungi in the air supply system of auditorium (Sabourand Agar).

- Most of the bacterial isolates from the auditorium and supply ducts air belonged to genera *Micrococcus* and *Bacillus* but *Staphylococcus* and *Actinomyces* were also present.
- Among fungi isolated from both outdoor and indoor air the genera: *Cladosporium*, *Aspergillus* and *Penicillium*, dominated.

ACKNOWLEDGEMENTS

This study was supported by project PB 4955/T02/2008/34, financed by the Polish Ministry of Science and Higher Education.

REFERENCES

Albrecht A., Kiel K. & Kolk A. 2007. Strategies and Methods for Investigation of Airborne Biological Agents From Work Environments in Germany. *International Journal of Occupational Safety and Ergonomics (JOSE)* 13 (2): 201–213.

Crook B. & Burton N.C. 2010. Indoor moulds, Sick Building Syndrome and building related illness. *Fungal Biology Reviews*, doi:10.1016/j.fbr.2010.05.001 (article in press).

Crook B. & Sherwood-Higham J.L. 1997. Sampling and assay of bioaerosols in the work environment. *J Aerosol Sci.* 28(3): 417–426.

Directive 2000/54/EC of the European Parliament and of the Council of 18 September 2000 on the protection of workers from risks related to exposure to biological agents at work. *Official Journal of the European Communities* 2000; L262: 21–45.

European Committee for Standardization (CEN). Workplace atmospheres—guidelines for measurement of airborne microorganisms and endotoxins (Standard No. 13098:2000). Brussels, Belgium: CEN; 2000.

European Committee for Standardization (CEN). Workplace atmospheres—determination of airborne endotoxins (Standard No. 14031:2003). Brussels, Belgium: CEN; 2003.

Górny R. 2004. Biologiczne czynniki szkodliwe: normy, zalecenia i propozycje wartosci dopuszczalnych. *Podstawy i Metody Oceny Srodowiska Pracy* 3(41): 17–39 (in Polish).

Górny R.L. & Dutkiewicz J. 2002. Bacterial and fungal aerosols in indoor environment in Central and Eastern European countries. *Ann. Agric. Environ. Med.* 9(1): 17–23.

Kim K.Y. & Kim C.N. 2007. Airborne microbiological characteristics in public buildings of Korea. *Building and Environment* 42: 2188–2196.

Lee T., Grinshpun S.A., Martuzevicius D., Adhikari A., Crawford C.M., Lou J. & Reponen T. 2006. Relationship between indoor and outdoor bioaerosols collected with a button inhalable aerosol sampler in urban homes. *Indoor Air* 16: 37–47.

Lis D.O. & Pastuszka J.S. 1997. Bacterial and fungal aerosol in homes without and with mold problem, in Upper Silesia, Poland. Preliminary results. *J. Aerosol Sci.* 28(1): 665–666.

Medrela-Kuder E. 2003. Seasonal variations in the occurrence of culturable airborne fungi in outdoor and indoor air in Cracow. *International Biodeterioration & Biodegradation* 52: 203–205.

Meklin T., Reponen T., Toivola M., Koponen V., Husman T., Hyvärinen A. & Nevalainen A. 2002. Size distributions of airborne microbes in moisture-demaged and reference school buildings of two constructions types. *Atmospheric Environment* 36: 6031–6039.

Nevalainen A. & Seuri M. 2005. Of microbes and men. *Indoor Air* 15(Suppl 9): 58–64.

NF-S-90-351, Procédures de Réception et de Controlle des Salles d'Opération. Qualité de l'Air, France, 1987.

NHS Estates. Health Technical Memorandum 2025. National Health Service. Ventilation in Healthcare Premises. Management Policy, 1994.

Palmgren U., Strom G., Blomquist G. & Malmberg P. 1986. Collection of airborne microorganisms on Nucleopore filters, estimation and analysis – CAMNEA method. *J. Appl. Bacteriol.* 61: 401–406.

Parat S., Perdrix A., Fricker-Hidalgo H., Saude I., Grillot R. & Baconnier P. 1997. Multivariate analysis comparing microbial air content of an air-conditioned building and a naturally ventilated building over one year. *Atmospheric Environment.* 31(3): 441–449.

Pasanen A.L., Pasanen P., Jantunen M.L. & Kalliokoski P., 1991. Significance of air humidity and air velocity for fungal spore release into the air. *Atmospheric Environment* 25A (2): 459–462.

Reponen T., Gazenko S.V., Grinshpun S.A., Willeke K. & Cole E.C. 1998. Characteristics of airborne actinomycete spores. *Appl. Environ. Microbiol.* 64(10): 3807–3812.

Reponen T., Willeke K., Ulevicius V., Reponen A. & Grinshpun S.A. 1996. Effect of relative humidity on the aerodynamic diameter and respiratory deposition of fungal spores. *Atmospheric Environment* 30(23): 3967–3974.
Ross M.A., Curtis L., Scheff P.A., Hryhorczuk D.O., Ramakrishnan V., Wadden R.A. & Persky V.W. 2000. Association of asthma symptoms and severity with indoor bioaerosols. *Allergy* 55: 705–711.
Sattar S.A. & Ijaz M.K. 1987. Spread of viral infections by aerosols. *CRC Crit. Rev. Environ. Control* 17: 89–131.
Shoemaker R.C. & House D.E. 2005. A time-series study of sick building syndrome: chronic, biotoxin-associated illness from exposure to water-demaged buildings. *Neurotoxicology and Teratology* 27: 29–46.
Stetzenbach L.D. 1998. Microorganisms and Indoor Air Quality. *Clinical Microbiology Newsletter* 20(19): 157–161.
Tham K.W. & Zuraimi M.S. 2005. Size relationship between airborne viable bacteria and particles in a controlled indoor environment study. *Indoor Air* 15 (Suppl 9): 48–57.
Toivola M., Alm S., Reponen T., Kolari S. & Nevalainen A. 2002. Personal exposures and microenvironmental concentrations of particles and bioaerosols. *J. Environ. Monit.* 4: 166–174.
Wang Y.F., Wang C.H. & Hsu K.L. 2010. Size and seasonal distributions of airborne bioaerosols in commuting trains. *Atmospheric Environment* 44: 4331–4338.

Management of Indoor Air Quality – Dudzińska (ed)
© 2011 Taylor & Francis Group, London, ISBN 978-0-415-67266-5

Aerosol particle concentrations and indoor air parameters in school classrooms

B. Połednik
Faculty of Environmental Engineering, Lublin University of Technology, Lublin, Poland

ABSTRACT: The objective of the study was to measure the mass concentrations of aerosol particles in classrooms and to determine their associations with number and activities of present students, indoor thermal conditions (temperature, relative humidity), CO_2 concentration and season of the year. Monitoring took place in 6 classrooms in 3 schools when the classrooms were occupied and they were vacant during the heating and summer seasons. Particle Mass (PM) concentrations were measured by laser photometer. The average PM concentrations ranged from 185 to 363 $\mu g/m^3$ and from 132 to 305 $\mu g/m^3$ in the heating and summer season, respectively. The highest ratio of PM concentrations in occupied and vacant classrooms (O/V) was for coarse particle fractions. Due to the fact that measurements were carried out in real field conditions, highly significant correlations between the PM concentration and the number of students present were not found. Higher coarse particle fraction concentrations were observed at higher indoor air temperatures and CO_2 concentrations. Concentrations of small particle fractions were positively correlated with indoor air relative humidity.

Keywords: PM concentration, particular matter, indoor air, classroom environment

1 INTRODUCTION

People generate significant amounts of indoor aerosols, which may have an important impact on short-term exposures (Jankowska et al. 2003, Nazaroff 2004, Afshari et al. 2005). Comparative risk studies have constantly ranked indoor aerosols among the top environmental risks to public health (EUR 19529/EN 2000, Schneider et al. 2003, Mølhave et al. 2004, WHO 2007).

Young people are at greater risk because of the hours spent in school facilities and because young people are more susceptible than adults to the harmful effects of indoor aerosols (Blondeau et al. 2005, Weichenthal et al. 2007). Particles which are suspended in classroom air are significantly responsible for increased allergies, respiratory infections and asthma exacerbations. In turn, such health problems are related to reduced student attendance and performance (Steinemann 2004, Mendell & Heath 2005).

Despite designated smoking areas (nowadays seldom in public spaces) and designated places for open fire (kitchen with gas stoves, etc.), education facilities are among the indoor premises suspected of high aerosol concentrations. The probable high concentrations are mainly due to high occupation per square meter, and to increased movement of the students during breaks and when changing classrooms.

Therefore, interest in classroom aerosols and their relation to the outdoor and indoor air conditions is increasing.

Generally it is accepted, that the presence of students and their physical activity are the main factors which influence the aerosol concentration in classrooms (Fromme et al. 2007, Morawska et al. 2009, Połednik & Dudzińska 2010). The microclimatic changes in classrooms are also significant (Braniš et al. 2005, Polednik et al. 2007). In spite of intensive research it is still not entirely clear how the concentration of particles is associated with the number and activities of students and indoor air parameters in classrooms. Information concerning the factors which influence the

Table 1. Characteristics of monitored classrooms.

Characteristic	School 1		School 2		School 3	
	Computer science classroom C1(1)	Biology classroom C2(1)	Computer science classroom C1(2)	Biology classroom C2(2)	Computer science classroom C1(3)	Biology classroom C2(3)
Floor area [m^2]	29	46	57	69	71	59
Volume [m^3]	96	130	205	250	221	183
Floor number	2	1	3	2	3	1
Floor material	ceramic	ceramic	ceramic	ceramic	ceramic	ceramic
Windows type	PCV	PCV	PCV	PCV	PCV	PCV
Blackboard type	white	black	white	black	white	black
Number of students max.	18	32	18	34	22	32
Heating season measurements	O 03.09.10 V 03.14.10	O 03.10.10 V 03.13.10	O 03.30.10 V 04.02.10	O 03.31.10 V 04.03.10	– –	– –
Summer season measurements	O 06.10.10 V 06.12.10	O 06.11.10 V 06.13.10	O 05.20.10 V 05.22.10	O 05.21.10 V 05.23.10	O 04.20.10 V 04.24.10	O 04.21.10 V 04.25.10

O – occupied classroom
V – vacant classroom

changes in aerosol particle concentrations, and monitoring of these concentrations in classrooms, is important for determining the potential health effects on students.

Aerosol concentrations are not among the parameters widely studied during routine inspections of indoor air quality in schools and other education facilities. The aerosol concentration levels have been examined in a number of schools in Lublin and its vicinity. The study has been conducted on the occasion of the indoor air quality assessment in schools and the determination of its impact on the students' health.

This paper presents the preliminary results of aerosol Particle Mass (PM) concentration measurements in selected schools.

2 SITE AND METHODS

The experiments were carried out in the classrooms of three schools in Lublin, Poland. Monitoring was done both while rooms were being heated in the heating season, and then again in the summer season. The schools were located in different parts of the city in densely populated residential areas, away from busy streets. No specific sources of emissions were located in the vicinity. One of the schools was built in 1930, the remaining two were built in 1970. Table 1 summarizes the characteristics of the monitored classrooms, labeled S1/S2(1) – S1/S2(3), along with the measurement dates. In each school (numbers in parentheses), two classrooms were chosen for monitoring. The classrooms were used for the purposes of teaching either computer science or biology/physics and they were located on the first, second and third floor of the school buildings. The classrooms had various floor areas (from 29 to 71 m^2) and volumes (from 96 to 250 m^3). The maximum occupancy depended on the size of the classrooms and ranged between 18–34 students. The classrooms had ceramic floors and were equipped with standard school furniture. The traditional blackboards for chalk and whiteboards for color-board markers were used in the classrooms. All the monitored classrooms relied on natural ventilation through large double-glazed windows.

The measurements in each classroom were made during teaching hours and lasted for a few hours per day. Additionally, background measurements were conducted in the vacant classrooms.

The measurement equipment was placed on empty desks in the middle of the classroom. The sampling devices were located at a height of 110 cm. Aerosol particle mass concentrations were continuously determined by means of the TSI Dust Trak DRX model 8533 (TSI Inc., Shoreview, MN, USA). This laser photometer can simultaneously measure size-segregated mass fraction concentrations corresponding to PM_1, $PM_{2.5}$, Respirable, PM_{10}, and TOTAL particle matter size fractions. The measured particle size range is from 0.1 to 15 μm, at a concentration level of between 0.001 and 150 mg/m^3. The Onset Hobo U12 External Data Logger sensors were used to monitor the indoor air temperature and Relative Humidity (RH). The temperature measurement range of the sensors was from −20°C to 70°C and the RH range from 5 to 95%. The GMD/W20 sensor manufactured by VAISALA with the measurement range from 0 to 5 000 ppm, was used to measure the CO_2 concentration. The chosen sampling interval for all the measuring instruments used was 60 s. Before measurements, the instruments were calibrated by their producers. The distribution of measurement results allowed for the use of Spearmen correlation coefficients as indicators of the association between the PM concentrations, indoor air parameters, occupation and characteristic parameters of classrooms. The differences between the results obtained in the heating and summer seasons were calculated for classrooms monitored during those seasons.

3 RESULTS AND DISCUSSION

3.1 *Aerosol particle mass concentrations*

Many coexisting factors influence the aerosol particle mass concentration levels in the monitored classrooms, especially when students are present. The exemplary time series of the measured PM fraction concentrations in an occupied biology classroom C2(2), in one of the examined schools in the heating (I) and summer (II) season, are shown in Figure 1. The time series data for indoor air temperature, relative humidity and CO_2 concentration are also presented. For comparison, Figure 2 presents the time series of the same aerosol PM fractions and indoor air parameters when this classroom was vacant. The presence of students in the classroom is an important factor which changes all the measured parameters. The observed parameter fluctuations seem to be strongly correlated with the occupancy. The maximum PM concentrations of the TOTAL fraction were up to twelve times higher than the baseline indoor levels during the classroom occupancy (100 μg/m^3), and up to fifteen times higher than the baseline levels during classroom vacancy (~80 μg/m^3).

The highest increase in these particle concentrations was observed just as classes began, and during breaks. This can be explained by the increased physical activity of students during that time; students entering the classroom, taking their seats and leaving the classroom, respectively. During this time of activity, an increase in air circulation can be observed in the classroom. The intensification of the air movement causes e.g. increased resuspension of dust particles deposited on surfaces in the classroom. This resuspension significantly contributes to the increased aerosol particle concentration in the indoor air (Nazaroff 2004, Mullen et al. 2011). During the breaks between classes, especially in the summer season, the classroom was ventilated by opening the windows and doors. This ventilation caused additional intensive indoor air movement and at the same time caused substantial changes in the indoor aerosol PM concentrations.

During the classes, when the students had already taken their seats and their activity was rather moderate, a decrease in the PM concentrations was observed. The concentrations of coarse particles, characterized by PM_{10} and TOTAL fractions, decreased by several times. The reason for this is because these particles mainly undergo gravitational sedimentation and are relatively quickly removed from the indoor air and deposited on indoor surfaces as dust (Nazaroff 2004). Concentrations of fine aerosol particles (PM_1 and $PM_{2.5}$ fractions) decreased to a lesser extent. This in turn results from the fact that relatively fewer of these particles were introduced into the air during the class exchange breaks, and because such aerosols are characterized by greater stability. Fine particles undergo slower sedimentation. In addition, the processes which retain these fine particles in suspension e.g. electrostatic repulsions or Brown motions are more effective.

In the C2(2) classroom (Figs 1–2) but also in almost all the monitored classrooms in the heating season when the students were present, the mean concentrations of coarse particles were higher than in the summer season. In the heating season, it may be that a larger number of coarse aerosol particles are brought in on the students' winter clothes and shoes and can be more easily resuspended and introduced to the indoor air (Braniš at al. 2005).

In the heating season, the average concentrations of the TOTAL fraction during the periods of student occupancy ranged from 185 μg/m^3 in C2(2) to 363 μg/m^3 in C2(1). The average across the investigated four classrooms was 280 μg/m^3. In the summer season, the minimum average concentration of this fraction amounted to 132 μg/m^3 in C1(1). The maximum of 305 μg/m^3 was

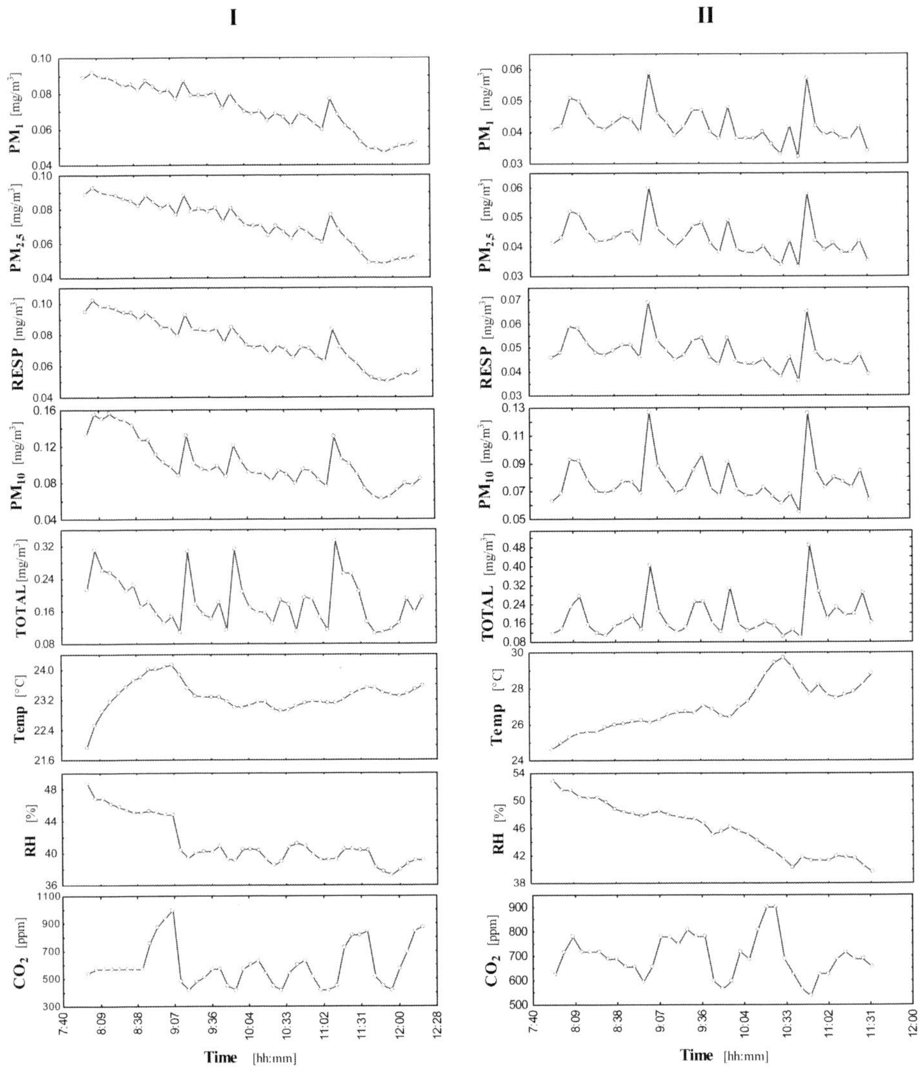

Figure 1. Time series of Particle Matter (PM) concentrations and indoor air temperature, Relative Humidity (RH) and CO_2 concentration in the occupied biology classroom C2(2) in one of the examined schools in the heating (I) and summer (II) season.

measured in C1(2). The average across all six classrooms amounted to 211 $\mu g/m^3$. On average, the particle mass concentrations decreased by 21% (PM_1), 28% ($PM_{2.5}$), 18% (PM_{10}) and 24% (TOTAL) in the summer compared to the heating season. These differences were statistically significant ($p < 0.05$).

The ratios of the average mass concentrations in the occupied and vacant classrooms (O/V) in the heating and summer seasons are reported in Table 2. In both seasons the highest average O/V ratios were for the coarse aerosol particle fractions (TOTAL and PM_{10}). For the TOTAL fraction values of O/V ratios in the heating season were in the range between 3.30 in C2(2) and 5.03 in C1(1) (average for four classrooms = 4.02) and in the summer season varied from 1.52 in C1(1) to 3.72 in C2(3) (average for six classrooms = 2.69). It is necessary to point out, that the number of students varied in the occupied, monitored classrooms (Tab. 2).

Because a parallel collection of outdoor and indoor PM concentrations was unworkable, the possible influence of outdoor to indoor conditions can only be estimated. It is important to note that our unpublished results of outdoor PM concentration measurements performed in Lublin

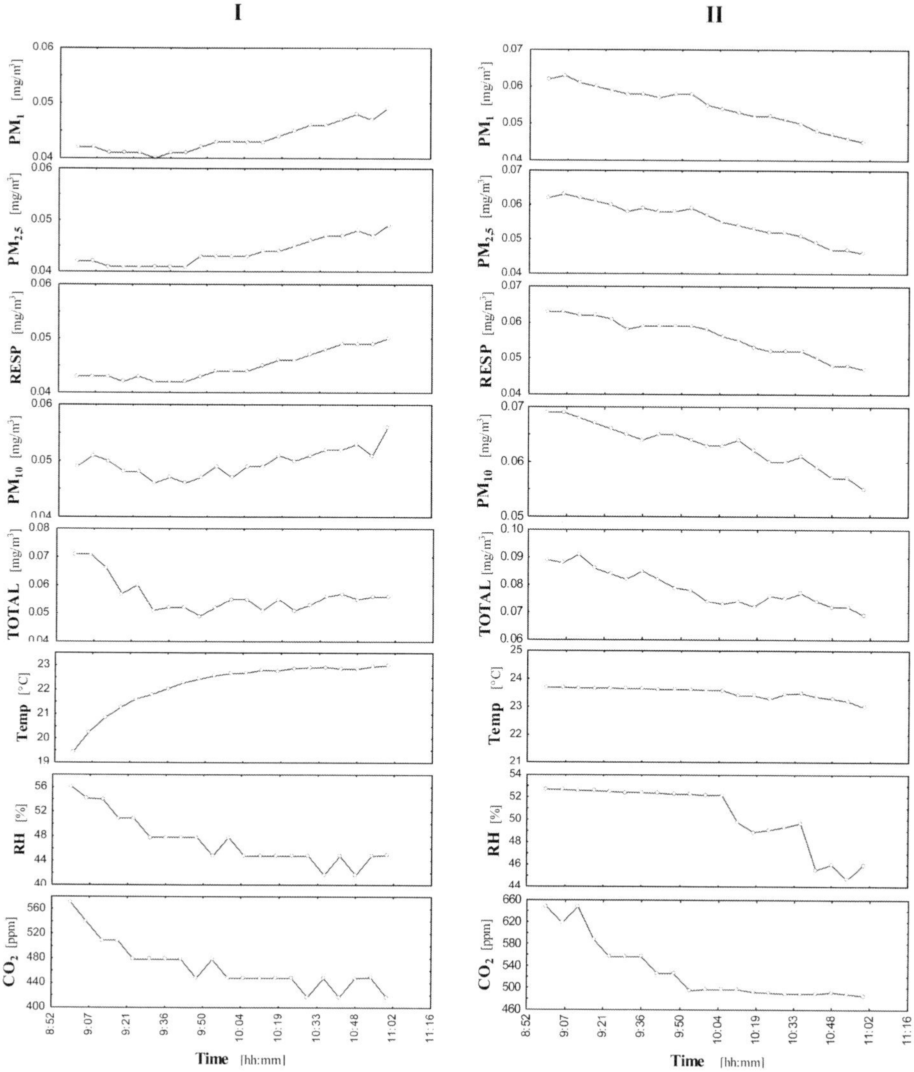

Figure 2. Time series of Particle Matter (PM) concentrations and indoor air temperature, Relative Humidity (RH) and CO_2 concentration in the vacant biology classroom C2(2) in one of the examined schools in the heating (I) and summer (II) season.

during the winter and summer are much lower than the concentrations observed in the occupied classrooms, especially with respect to the average coarse particle levels.

3.2 *Indoor air parameters*

From the performed measurements, changes of indoor air parameters in the investigated classrooms can be seen to be dependent on the occupation and season. During the time when classes were being held, an increase in the temperature by several degrees and in the RH by several percent were observed, however, such increases were usually smaller in the heating season. In the summer season, the impact of the outdoor air thermal parameters seems to be more visible. Measurements performed in vacant classrooms confirm this impact. When the students were present, the CO_2 concentration was also subject to high fluctuations. During the classes, the CO_2 concentration increased and then exponentially decreased during the classroom breaks. In the summer, as a result

Table 2. Occupied to vacant ratios (O/V) of PM concentrations in the monitored classrooms in the heating (I) and summer (II) season.

Classroom	Season	PM_1	$PM_{2.5}$	PM_{10}	TOTAL	N
C1(1)	I	2.14	2.09	2.69	5.03	13.2 ± 6.9 (16–18)
	II	1.74	1.76	1.63	1.52	11.6 ± 5.6 (15–16)
C2(1)	I	1.90	1.92	2.65	4.27	9.2 ± 4.6 (10–13)
	II	1.50	1.50	1.68	2.34	6.5 ± 7.5 (7–17)
C1(2)	I	0.75	0.75	1.36	3.49	11.2 ± 5.9 (12–17)
	II	1.24	1.27	1.96	3.59	11.2 ± 5.3 (12–16)
C2(2)	I	1.61	1.64	2.04	3.30	18.4 ±10.2 (22–29)
	II	0.76	0.78	1.22	2.45	19.2 ± 9.7 (25–26)
C1(3)	II	0.52	0.53	1.14	2.49	10.7 ± 8.8 (12–24)
C2(3)	II	1.21	1.22	1.94	3.72	10.5 ± 10.9 (14–28)

N – mean ± sdev. (median-maximal) number of students

of the changes in ventilation in the classrooms (e.g. frequent opening of windows), less regular changes in the CO_2 level were observed.

In almost all monitored classrooms, where classes were being held, the indoor CO_2 levels exceeded 1000 ppm, which according to the ASHRAE Standards 62-1989 is the threshold concentration for satisfactory comfort. In the heating season, the CO_2 concentration in 56% of the teaching hours and in the summer season in 41% of the teaching hours did not meet the ASHRAE Standards. CO_2 concentration readings for over 1500 ppm were found during 25% and 7% of the teaching hours, in the heating and summer seasons, respectively.

In vacant classrooms, the measured CO_2 level was also not stable. The initially registered higher values of the CO_2 concentration could possibly be the result of the room being previously occupied as well as the result of the presence of the persons setting up the measurement devices.

Measurements were carried out in real field conditions with student activity, classroom ventilation, and outdoor air parameters all changing. For this reason, the highly significant correlations between the aerosol concentration and the number of present students or indoor air parameters were hard to find. Strong correlations between the PM concentration, classroom location in school buildings, classroom volume, floor area, finished materials and blackboard or whiteboard used in the classroom were also not observed. The best exemplary associations that could be drawn between the measured concentration of the TOTAL aerosol particle fraction and the indoor air temperature and the CO_2 concentration in the monitored occupied classrooms, in the heating and summer seasons, are shown in Figure 3. The relation between the concentration of fine fraction PM_1 and the indoor air relative humidity is also presented.

There was an observable increase in the concentration of almost all the measured aerosol particle fractions when the indoor air temperature and CO_2 concentration rose. The PM_1 concentrations were also higher when the RH of the indoor air increased. Similar trends were observed in almost all of the investigated classrooms when the students were present. These trends were more distinct in the summer season.

3.3 *Comparison with previous research*

The presented results show that the exposure to aerosol particles in the investigated classrooms during the teaching hours is high. Similar results were obtained in previous studies. Fromme et al. (2007) reported the median daily values of PM concentrations measured in 58 classrooms in Munich to be between 2.7 and 80.8 $\mu g/m^3$ for $PM_{2.5}$ and between 16.3 and 313 $\mu g/m^3$ for PM_{10}, in winter. In summer, concentrations ranged between 4.6 and 34.8 $\mu g/m^3$ ($PM_{2.5}$) and between 18.3 and 178 $\mu g/m^3$ (PM_{10}). Based on the measurements performed in two classrooms in one primary school in Munich, Fromme et al. (2008) reported median $PM_{2.5}$ concentrations of 37.4 $\mu g/m^3$ and PM_{10} of 118.2 $\mu g/m^3$. Lower PM concentrations in classroom were observed by Braniš et al. (2005). They found that average indoor daytime concentrations were 21.9 $\mu g/m^3$ for $PM_{2.5}$ and 42.3 $\mu g/m^3$ for PM_{10}. However, it has to be pointed out that the PM concentrations were measured by the gravimetrical method. The readings from this method, as was noted by the researches, are

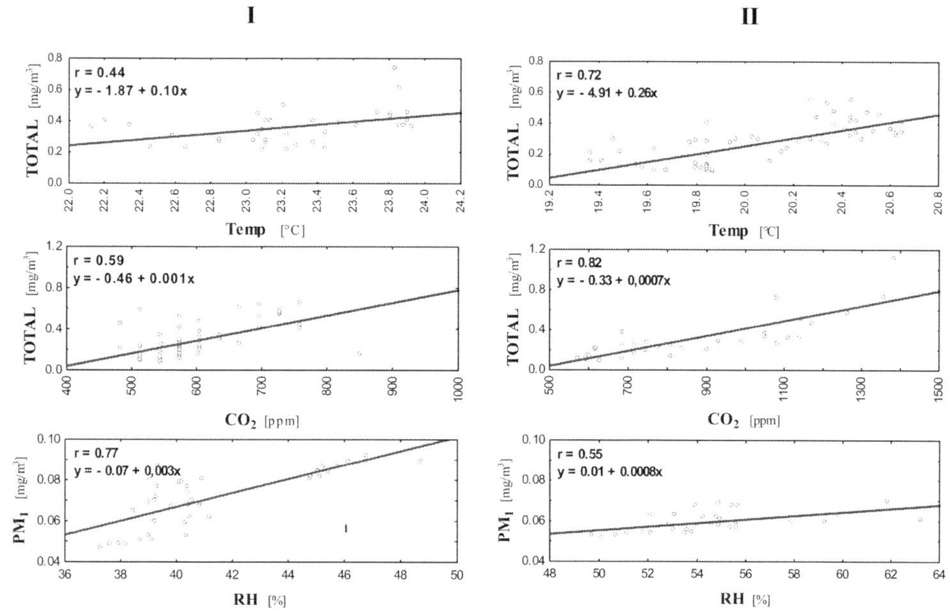

Figure 3. Relation between mass concentration of TOTAL fraction and indoor air temperature and CO_2 concentration, and between mass concentration of PM_1 fraction and indoor air relative humidity in the occupied classrooms in the heating (I) and summer (II) season.

lower than when using Dust Trak photometer readings. Relatively small amounts of PM_{10} were found in 40 classrooms in Korea (Son et al. 2006). In summer, the Korean classrooms had a mean concentration of 46.7 $\mu g/m^3$ and in winter 39.1 $\mu g/m^3$. The relationship between PM concentration in classrooms and the season of the year in which measurements were performed are mostly the same as were noted in previous research. This means that concentrations are higher in the heating season (winter) than in summer. The observed differences are most likely due to the different ways used to ventilate the rooms in these seasons. In winter, due to decreased ventilation, the PM concentrations are strongly influenced by the indoor activities of students. In summer, PM concentrations are very dependent on outdoor levels of aerosol particles (Fromme et al. 2007).

In our study, the monitored CO_2 concentrations in the occupied classrooms were consistent with the results from other published studies (Fromme et al. 2007, Heudorf et al. 2009). Very high CO_2 levels (above 1500 ppm) were more often found in the heating season, indicating insufficient ventilation in the classrooms.

The impact of indoor air thermal parameters is emphasized in the paper. As in other research studies (Braniš et al. 2005, Połednik et al. 2007, Połednik & Dudzińska 2008), we found high positive associations recorded between finer fractions and indoor air relative humidity. With some exceptions significant correlations were also found between indoor air temperature and PM concentrations made up of mainly coarse fractions.

It was not possible to demonstrate a dependency between PM concentrations and factors related to the characteristics of the monitored schools and classrooms, namely location, room size, furniture and equipment. The expected statistically significant correlations with actual number of students present in the classrooms were also not obtained, even though the presence of students and their activities represent a significant source of aerosol particle coarse fractions (Braniš et al. 2005, Fromme et al. 2007). The lack of correlations presumably results from a limited set of available data for analysis.

4 CONCLUSIONS

The obtained measurement results show that the exposure to aerosol particles is high in the monitored classrooms. The particle mass concentration levels, in all the monitored classrooms, are mainly dependent on the presence and activity of the students as well as on the ventilation conditions and the season.

Higher coarse particle concentrations in most of the classrooms are accompanied by higher indoor air temperatures and higher CO_2 concentrations. Similarly, the PM_1 concentrations are associated with indoor relative air humidity.

Therefore, research on simple and efficient methods to decrease and control PM levels in school classrooms and thereby reduce the students' risk of exposure, should be undertaken.

ACKNOWLEDGEMENTS

This research was carried out within the Projects 4955/T02/2008/34, financed by the Polish Ministry of Science and Higher Education.

REFERENCES

Afshari, A., Matson, U. & Ekberg, L.E. 2005. Characterization of indoor sources of fine and ultrafine particles: a study conducted in a full-scale chamber. *Indoor Air* 15(2): 141–150.

Blondeau, P., Iordache, V., Poupard, O., Genin, D. & Allard F. 2005. Relationship between outdoor and indoor air quality in eight French schools. *Indoor Air* 15: 2–12.

Braniš, M., Řezáčová, P. & Domasová M. 2005. The effect of outdoor air and indoor human activity on mass concentrations of PM_{10}, $PM_{2.5}$ and PM_1 in a classroom. *Environ. Research* 99: 143–149.

Fromme, H., Twardella, D., Dietrich, S., Heitmann, D., Schierl, R., Liebl, B. & Ruden, H. 2007. Particulate matter in the indoor air of classrooms – exploratory results from Munich and surrounding area. *Atmospheric Environment* 41(4): 854–866.

Heudorf, U., Neitzert, V. & Spark, J. 2009. Particulate matter and carbon dioxide in classrooms – The impact of cleaning and ventilation. *Int J Hyg Environ Health* 212(1): 45–55.

Jankowska, E., Kondej, D. & Jankowski, T. 2003. Particle size distribution and aerosol concentration at office workplaces. *Abstracts of the European Aerosol Conference* 1: 391–392.

Mendell, M.J. & Heath, G.A. 2005. Do indoor pollutants and thermal conditions in schools influence student performance? A critical review of the literature.*Indoor Air* 15(1): 27–52.

Mølhave, L., Kjærgaard, S.K. & Attermann, J. 2004. Respiratory effects of experimental exposure to office dust. *Indoor Air* 14(6): 376–384.

Morawska, L., He, C., Johnson, G., Guo, H., Uhde, E. & Ayoko, G. 2009. Ultrafine Particles in Indoor Air of School: Possible Role of Secondary Organic Aerosols. *Environmental Science & Technology* 43(24): 9103–9109.

Mullen, N.A., Bhangar, S., Hering, S.V., Kreisberg, N.M. & Nazaroff, W.W. 2011. Ultrafine particle concentrations and exposures in six elementary school classrooms in northern California. *Indoor Air* 21: 77–87.

Nazaroff, W.W. 2004. Indoor particle dynamics. *Indoor Air* 14: 175–183.

Połednik B. & Dudzińska M.R. 2008. Occupation, thermal condition and aerosol particle number concentration in a lecture room. XII International Conference on Air Conditioning Air Protection & District Heating, Wroclaw, 439–444.

Połednik, B. & Dudzińska, M.R. 2010. Ventilation control based on the CO_2 and aerosol concentration and perceived air quality measurements – a case study. *Archives of Environmental Protection* 36: 67–80.

Połednik, B., Raczkowski, A. & Dumała S. 2007. Aerosol particle concentration and the thermal conditions in a lecture room. *Environmental Engineering* – Pawlowski, L., Dudzinska, M. & Pawlowski A. (eds), Taylor & Francis Group, London, 387–390.

Schneider, T., Sundell, J., Bischof, W., Bohgard, M., Cherrie, J.W., Clausen, P.A., Dreborg, S., Kildesø, J., Kjærgaard, S.K., Løvik, M., Pasanen, P. & Skyberg, K. 2003. 'EUROPART'. Airborne particles in the indoor environment. A European interdisciplinary review of scientific evidence on associations between exposure to particles in buildings and health effects. *Indoor Air* 13(1): 38–48.

Son, B.S., Song, M.R. & Yang, W.H. 2006. A study on PM_{10} and VOCs concentrations of indoor environment in school and recognition of indoor air quality. Proceedings of Indoor Air 2005: 827–832.

Steinemann, A. 2004. Human exposure, health hazards, and environmental regulations. *Environmental Impact Assessment Review* 24(7–8): 695–710.

Weichenthal, S., Dufresene, A. & Infante-Rivard C. 2007. Indoor ultra-fine particles and childhood asthma: exploring a potential concern. *Indoor Air* 17: 81–91.

WHO 2007. Indoor Air Pollution: National Burden of Disease Estimates, Geneva, Switzerland.

Management of Indoor Air Quality – Dudzińska (ed)
© 2011 Taylor & Francis Group, London, ISBN 978-0-415-67266-5

Indoor activity of ^{222}Rn in different types of rooms in domestic dwellings

A. Dołhańczuk-Śródka, Z. Ziembik & M. Wacławek
Independent Chair of Biotechnology and Molecular Biology, Opole University, Opole, Poland

ABSTRACT: In this work monitoring of ^{222}Rn was carried out in different rooms of domestic dwellings located in Opole (south-western Poland). Passive detectors (LR-115 type II films; Kodak, Pathe) were used in the survey. The detectors were placed in kitchens, bathrooms, cellars and living rooms and remained there from the end of November to the end of February 2008. The results showed that in all of the rooms examined the ^{222}Rn activity slightly exceeded 135 Bq/m^3. This value is lower than the European Commission recommendations for existing buildings. No significant differences in ^{222}Rn activity in the investigated rooms was found, except in kitchens and bathrooms where activities were somewhat different.

Keywords: indoor radioactivity, ^{222}Rn, passive detectors

1 INTRODUCTION

The short-lived decay products of radon are the most important contributors to human exposure to radioactivity from natural sources. About one third of the total radiation dose to humans is due to inhalation of short-lived radon progeny in the indoor atmosphere. In 2009 an average Polish citizen received a dose of 3.19 mSv, in which the contribution of radon was 37.6% (PAA Report 2010).

The activity concentrations of radon and their progeny are largely influenced by factors such as topography, type of house construction, building materials, temperature, pressure, humidity, ventilation, wind speed, and even the life style of the house residents (Jonassen 1975, Martz et al. 1991, Nazaroffand & Doyle 1985, Ramola et al. 1992, 1998, 2000, Segovia & Cejudo 1984, Subba Ramu et al. 1988).

Knowledge of the radon levels in buildings is important in assessing population exposure. Radon in indoor spaces may originate from exhalation from rocks and soils around the building or from construction materials used in walls, floors, and ceilings. Indoor radon concentrations are almost always higher than outdoor concentrations. Once inside a building, the radon cannot easily escape and accumulates. The sealing of buildings to conserve energy reduces the intake of outside air and worsens the situation. Radon levels are generally highest in cellars and basements because these areas are nearest to the source and are usually poorly ventilated. Radon can seep out of the ground and build up in confined spaces, particularly underground, e.g. in basements of buildings, caves, mines etc, and ground floor buildings. High concentrations can also be found in buildings because they are usually at slightly lower pressure than the surrounding atmosphere and so tend to suck in radon (from the soil) through cracks or gaps in the floor (Abd El-Zaher & Fahmi 2008).

Studies of radon and its progeny in the air inside the homes carried out in many countries showed wide variations, from a few Bq/m^3 even to tens of thousands Bq/m^3 (UNSCEAR 2006).

The International Commission on Radiological Protection gives two reference values (for dwellings), above which action must be taken to reduce radon concentrations: 400 Bq/m^3 – for houses already built and 200 Bq/m^3 for newly built homes (European Commission recommendations 2001). A recent recommendation of International Commission on Radiological Protection for the reference level for radon gas in dwellings is 300 Bq/m^3 (ICRP 2009).

In this work the monitoring of ^{222}Rn was carried out in different rooms of houses in Opole (south-western Poland). Radioactivity caused by this gas was measured in bathrooms, cellars, kitchens and living-rooms. The passive detectors LR-115 type II films (Kodak, Pathe) were used in the survey. A similar survey was carried out in earlier work on an area located about 30 km south of Opole (Dołhańczuk-Śródka et al. 2006, Dołhańczuk-Śródka et al. 2008).

2 MATERIALS AND METHODS

In the present investigations the indoor ^{222}Rn concentration was studied in nine houses. All of the houses investigated were located in the north-west part of Opole. They were similar to each other and were more than 50 years old. Each of them was cellared and built of clay-bricks. They were single-family houses with kitchens, living-rooms and bathrooms located in separate rooms. The largest distance between buildings did not exceed 1 km.

In the present investigations, a passive technique was employed using Solid State Nuclear Track Detectors (SSNTDs) which are sensitive to alpha particles in the energy range of the particle emitted by radon. SSNTDs also have the advantage of being mostly unaffected by humidity, low temperatures, moderate heating and light.

LR-115 Type 2 (12 m cellulose nitrate on 100 m polyester base) plastic track detectors each with a size of about (1.5 cm × 2 cm) were used.

The detectors were placed in kitchens, bathrooms, cellars and living-rooms at a height of about 2–2.5 m above the floor and at the distance 15–25 cm from the wall. One detector was placed in each room. During three winter months (from the end of November to the end of February, 2008) the detectors remained in the rooms.

After 3 months of exposure, the detectors were subjected to chemical processing in a 10 M analytical grade sodium hydroxide solution at (60 ± 1)°C, for 90 min, in a constant temperature water bath to enlarge the latent tracks produced by alpha particles from the radon decay. After the etching, the detectors were washed for 30 minutes with running cold water, then with distilled water and finally with a 50% water/alcohol solution. After a few minutes of drying in air, the detectors were ready for track counting. The etched tracks were counted using an optical microscope at 40 × magnification, observing 0.059 cm^2 of detector surface. The number of tracks was counted for each plate in 30 microscope fields of view. The ^{222}Rn concentration a (in Bq/m^3) is calculated by the following formula:

$$a = (Q/t) \times 13.8 \quad (1)$$

where: Q – number of tracks per cm^2 plate; t – exposure time in days; 13.8 – empirical coefficient in $Bq \cdot d \cdot cm^2/m^3$, calculated from data published by Srivastava (Srivastava 1995).

Errors σ in the track densities were calculated by multiplying the track density by $(1/N)^{1/2}$, where N is the total number of tracks counted in a sample (El-Zaher & Fahmi 2008). In our measurements the background radiation from other than ^{222}Rn was not taken into account. However, the considerable similarity between the buildings investigated and the relatively low surface area on which they were located may indicate that the variability in track densities is a result of differences in the utilization of the rooms.

3 RESULTS AND DISCUSSION

For statistical computations the *R* language (R Development Core Team 2010) was used. *R* is a free software environment for statistical computing and graphics. The capabilities of *R* can be extended through packages, which allow specialized statistical techniques, graphical devices, programming interfaces and import/export capabilities to many external data formats.

It was expected that, in the winter season, the ^{222}Rn activity would be highest in the rooms, due to limited ventilation. In table 1 the values of minimum *min*, lower quartile q_l, median *ME*, upper quartile q_u, maximum *max*, arithmetic mean a_m and standard deviation *SD* of ^{222}Rn activity computed from the data are shown. The calculated σ error was high for low activities, reaching 100% for single tracks observed on detector surfaces are. For the highest activities σ was 14%.

Each one of the activities can be placed in a certain range of values with probability p. Figure 1 shows histogram of estimated probability p density (estimation of p changes with changes in a) of

Table 1. Statistical parameters computed from the ^{222}Rn activities.

min [Bq/m^3]	q_l [Bq/m^3]	*ME* [Bq/m^3]	q_u [Bq/m^3]	*max* [Bq/m^3]	a_m [Bq/m^3]	*SD* [Bq/m^3]
3	16	22	37	135	30	29

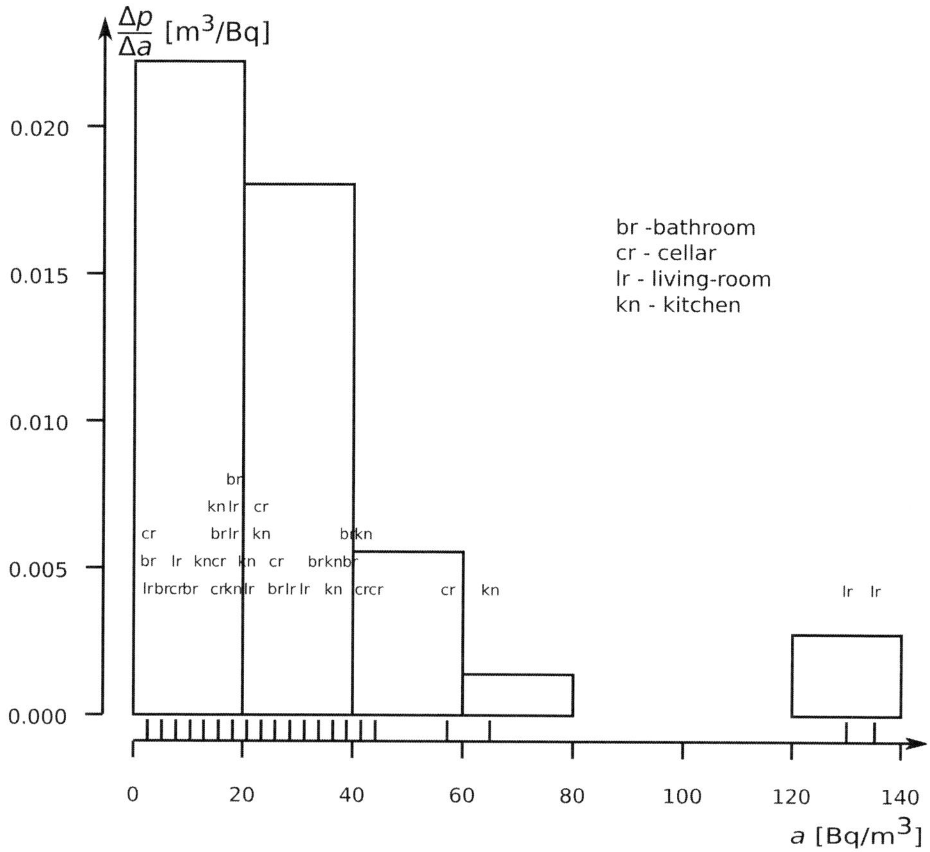

Figure 1. Histogram of density of ^{222}Rn activity in the rooms investigated.

^{222}Rn activity in the rooms investigated. In the abscissa the activity values are marked by vertical dashes and above each of them is located abbreviation indicating type of room. The histogram in Figure 1 show positively skewed, non-normal distribution of the data. The sequence of point marks does not pre-suppose any clear relationships between ^{222}Rn activities and room type. It can be assumed that the data distribution is common for all of the room types.

The differences between ^{222}Rn activities in different room types can be assessed by comparing the boxplots shown in Figure 2. In these plots the lower base of the rectangle is a lower quartile, the upper base is an upper quartile and a horizontal line dividing the rectangle represents a median. Whiskers are formed by connecting the formed box with short horizontal lines drawn for quantile $q = 0.95$ (upper whisker) and quantile 0.05 (lower whisker). The points marked by a circle are located more than 1.5 times of interquartile range from the median. The values of ^{222}Rn activities were similar in all of the rooms investigated. Median values are comparable and no significant differences in their values were observed. The ^{222}Rn activities considerably exceeded the common range only in one kitchen and two living-rooms.

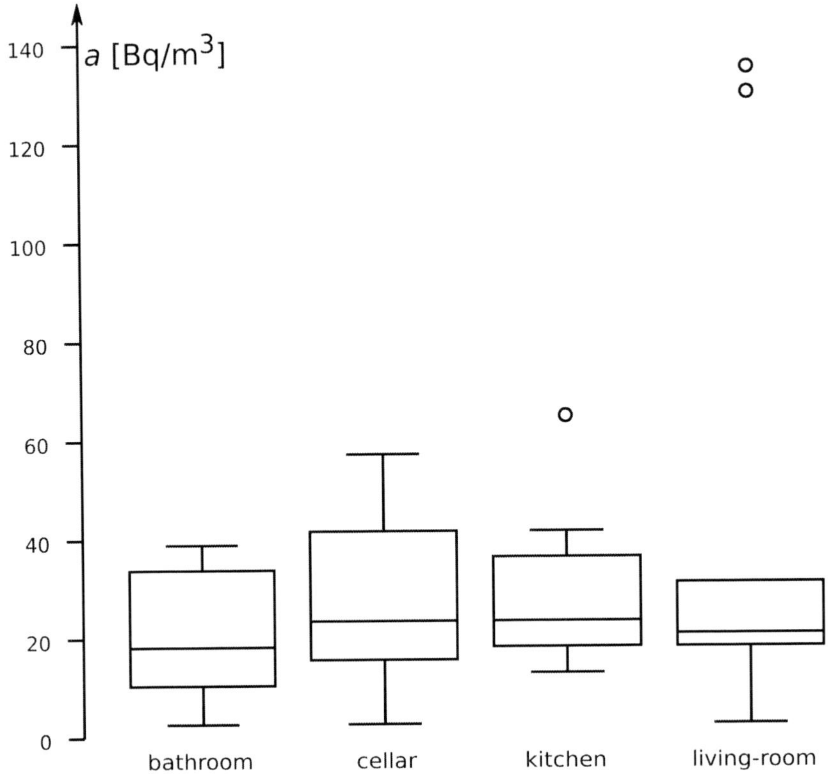

Figure 2. The differences between ^{222}Rn activities in different room types.

Activities of ^{222}Rn in different rooms were compared using statistical tests. Because of non-normal data distribution of variables the Mann-Whitney-Wilcoxon test was used. Pair-wise comparison showed no statistically significant differences between medians of activities in different rooms. The lowest p-value found was 0.29.

To enable utilization of statistical methods appropriate for the data with Gaussian (normal) distribution the ^{222}Rn activities were transformed using Box-Cox formulas (Venables & Ripley 2002). If the original random variable is x and the transformed one is x_t the Box-Cox formula can be expressed by the following relationships (Eqs 2a–b):

$$x_t = (x^{\lambda} - 1)/w \text{ for } \lambda \neq 0 \tag{2a}$$

$$x_t = \log x \text{ for } \lambda = 0 \tag{2b}$$

The estimated value of the λ exponent for our data was 0.166. After transformation, the normality of the data distribution was evaluated using Anderson-Darling, Lilliefors and Shapiro-Francia tests from the R package "nortest" (Gross 2006). The computed p-values were higher than the usually considered 0.05 limit, the smallest one of them was 0.12. These results indicated a normal distribution of the transformed data.

The mean transformed ^{222}Rn activity values computed for different room types were compared using the two-sided, paired t test. Equalities of activities in all possible pairs of rooms were assessed. Statistically significant differences in means were found only for bathroom and kitchen. Values of these parameters computed from the transformed data and recalculated back to the data with Bq/m^3 units were 17 and 27 in bathroom and kitchen respectively. Bearing in mind the bathroom/kitchen exception it could be concluded that within the natural variability in ^{222}Rn activity in the investigated rooms, the mean values of radon activities did not depend on room type in the houses investigated. However, differences might be larger in other seasons of the year.

Table 2. Ranges of ^{222}Rn activities in which the percentage of data is limited.

Percentage of all data [%]	from *a* [Bq/m^3]	*a* to [Bq/m^3]
68.3	8	52
95.5	2	108

Normality of the transformed activity distribution allowed the estimation of the common ^{222}Rn activities in the rooms.

The mean activity calculated from transformed data and then recalculated to original units was 22 Bq/m^3. The standard deviation calculated in the same way was 4 Bq/m^3. Table 2 show ranges and percentages of all data contained in two relevant activity ranges (8–52; 2–108).

Assuming the same distribution of ^{222}Rn activities in houses built on the area investigated, some predictions can be formulated.

The rooms investigated were in similar houses, located in a relatively small area. It can be supposed that the differences in track number on detectors were a result of differences in ^{222}Rn concentration in the rooms investigated.

All of the 27 measurement results were lower than the limits of ^{222}Rn activity applied in recommendations. However, these conclusions cannot be regarded as final. They were drawn from a relatively small data set and further investigation may change them. The crucial role in estimating a ^{222}Rn hazard can be assigned to the distribution function of radon activity in rooms. Application of more advanced statistical methods and collection of more extensive experimental data should augment and/or reinforce the final conclusions.

4 CONCLUSIONS

Our research shows that ^{222}Rn activities in bathrooms, cellars, kitchens and living-rooms were similar, the only statistically significant difference was found for bathrooms and kitchens. The distribution of activities was not normal, rather it was positively skewed. To draw correct inferences from statistical methods requiring normally distributed data, they have to be transformed using proper formulas. In the rooms investigated in this work the recommended limits of ^{222}Rn activity were not exceeded.

REFERENCES

Abd El-Zaher, M. & Fahmi, N.M. 2008. *Studying the Variation of Radon Level in Some Houses in Alexandria City, Egypt.* IX Radiation Physics & Protection Conference, 15–19 November 2008, Nasr City – Cairo: Egypt.

Dołhańczuk-Śródka, A., Chalecka, E., Smuda, M., Ziembik, Z. & Wacławek, M. 2006. Research of radon activity in the area of commune Strzeleczki. *Ecol. Chem. Eng.* 13(S4): 539–546 (in Polish).

Dołhańczuk-Śródka, A., Ziembik, Z., Chalecka, E. & Wacławek, M. 2008. Seasonal changes in ^{222}Rn activity in apartment rooms. *Ecol. Chem. Eng. S.* 5(1): 61–68 (in Polish).

European Commission recommendations 2001, 2001/928/Euratom. URL.

Gross, J. 2006. Nortest: Tests for Normality. R package version 1.0.

ICRP International Commission on Radiological Protection Statement on Radon, ICRP Ref 00/902/09 (November_2009).

Jonassen, N. 1975. On the effect of atmospheric pressure variations on the radon-222 concentration in unventilated rooms. *Health Phys.* 29: 216–220.

Martz, D.E., Rood, A.S., George, J.L., Pearson, M.D. & Langner Jr., H. 1991. Year to year variations in annual average indoor radon concentrations. *Health Phys.* 61: 413–437.

Nazaroff, W.W. & Doyle, S.M. 1985. Radon entry in the houses having crawl space. *Health Phys.* 48: 265–281.

PAA Report 2010. Działalność Prezesa Państwowej Agencji Atomistyki oraz ocena stanu bezpieczeństwa jądrowego i ochrony radiologicznej w Polsce w 2009 roku. Warszawa, 06.2010.

R Development Core Team (2010). R: A language and environment for statistical computing. R Foundation for Statistical Computing, Vienna, Austria. ISBN 3-900051-07-0.
Ramola, R.C., Kandari, M.S., Negi, M.S. & Choubey, V.M. 2000. A study of diurnal variation of indoor radon concentrations. *J. Health Phys.* 35: 211–216.
Ramola, R.C., Kandari, M.S., Rawat, R.B.S., Ramachandran, T.V. & Choubey, V.M. 1998. A study of seasonal variations of radon levels in different types of houses. *J. Environ. Radioact.* 39: 1–7.
Ramola, R.C., Singh, M., Singh, S. & Virk, H.S. 1992. Environmental radon studies using solid state nuclear track detectors.*J. Environ. Radioact.* 15: 95–102.
Segovia, N. & Cejudo, J. 1984. Radon measurements in the interior of household dwellings. *Nucl. Tracks Radiat. Meas.* 8: 407–410.
Srivastava, D.S., Singh, P., Rana, N.P.S., Naqvi A.H., Azam A., Ramachandran, T.V. & Subba Ramu, M.C. 1995. Calibration Factor for LR-115 Type II Track Detectors for Environmental Radon Measurements. *Nucl. Geophys.* 9(5): 487–495.
Subba Ramu, M.C., Shaikh, A.N., Murleedharan, T.S. & Ramachandran, T.V. 1988. A study of seasonal variation in the indoor concentrations of radon and its progeny. *Bull. Radiat. Prot.* 11: 81–88.
UNSCEAR 2006 Report Vol. II. Effects of Ionizing Radiation. Annex E: Sources-to-effects assessment for radon in homes and workplaces.
Venables, W.N. & Ripley B.D. 2002. *Modern Applied Statistics with S*, Springer.

Management of Indoor Air Quality – Dudzińska (ed)
© 2011 Taylor & Francis Group, London, ISBN 978-0-415-67266-5

Emission of Volatile Organic Compounds (VOC) from lacquered surfaces

M. Czajka & A. Dziewanowska-Pudliszak
Wood Technology Institute, Poznań, Poland

ABSTRACT: The level of emission of Volatile Organic Compounds (VOC) from acrylic, nitrocellulose, chemo-hardened and polyurethane lacquers applied to softwood (pine and larch) and hardwood species (oak, ash, beech and alder) was determined. Tests were carried out using the chamber method in accordance with ISO 16000 standards. The air from the chamber was sampled after 5, 24 and 48 hours of exposure using sorption tubes filled with sorbent (Tenax TA). Qualitative and quantitative analyses were done using TD-GC/MS technique. The tested wood species emitted volatile organic compounds in amounts ranging from 30 to 1500 $\mu g/m^3$. All tested lacquers showed a distinct effect of the wood species on the quantity and quality of emitted Volatile Organic Compounds (VOC). Lacquer coatings applied to a glass plate emitted the lowest VOC concentrations.

Keywords: emission of volatile organic compounds, chamber method, VOC, indoor air quality

1 INTRODUCTION

For centuries solid wood has been used for construction of houses, production of furniture and many other elements of furnishings. For the protection of such wood surfaces a wider and wider range of lacquers produced using cutting-edge technologies is used.

Bearing in mind that the quality of air in rooms intended for human habitation is especially important for health, the level of emission of volatile organic compounds from various wood materials found in the human environment has been widely studied, as well as the influence of wood preparation on VOC emission (Maupetit et al. 2002). Amongst the above-mentioned materials we find: flooring materials (Englund et al. 1996, Reiseri et al. 2002, Wilke et al. 2002, Wiglusz et al. 2002), materials used in the furniture industry (Gaca & Dziewanowska 2005), houses with wooden structure (Englund et al. 1996), builder's carpentry and joinery (Uhle 2010) and lacquers (Guo et al. 2002).

Studies of lacquer coatings are most often carried out on chemically neutral surfaces (e.g. glass). Such studies have demonstrated emission of lacquer thinners to the air containing a whole spectrum of volatile organic compounds depending on the coating type. On the other hand, tests of volatile organic compounds emission from "pure" pine wood and composite wood products showed that the analysed air contained 80% terpenes and the sum of volatile organic compounds (TVOC) was at the level of 920 $\mu g/m^3$ (Roffael 2006). Emission of volatile organic compounds from spruce wood tested after various exposure times resulted in TVOC emission from heartwood ranging from 3000 to 600 $\mu g/m^3$, and from sapwood ranging from 900 to 200 $\mu g/m^3$ (Ohlmeyer 2010). It was observed that emission of volatile compounds from wood and composite wood products may be influenced not only by the main wood components, such as cellulose, hemicelluloses and lignin, but above all by the following extraneous substances: resins, aromatic oils, tannins and dyes.

However, the manner of assessment and the level of emission of harmful chemical substances from defined combinations of lacquer and lacquered surface of solid wood is an open issue and there is no data on the influence of properties of different wood species on the emission of volatile organic compounds from lacquer coatings used for finishing.

Figure 1. Glass test chamber (0.225 m^3) for determine emission of VOC and formaldehyde.

The Wood Technology Institute carried out experiments aimed at identification of the effect of chosen softwood and hardwood species finished with various types of lacquer coatings on the level of volatile organic compounds emitted.

2 MATERIALS AND METHODS

Two species of soft wood, i.e. pine and larch, and four species of hardwood, i.e. oak, ash, beech and alder, were used for tests. Wood intended for tests was dried in a technical drying room. Before running the tests, samples were conditioned at 23°C and a relative humidity of 45%. After 2 weeks the moisture content of the test wood samples stabilised at the level of 9.2–0.7%.

Four types of lacquers most often used for wood surface finishing were selected for tests: water-based acrylic, nitrocellulose, chemo-hardened, and polyurethane.

Lacquer was applied in an amount equivalent to 130 g per m^2 of a sample. Lacquered samples were left in the open air for 24 hours which allowed drying of all types of tested coatings to the highest level required. The edges of all samples were sealed with a low-emitting aluminium tape.

All determinations of the VOC emission from the tested materials were performed by the chamber method according to ISO 16000-9. The glass test chamber with a volume of 0.225 m^3 was equipped with suitable accessories such as inlet and outlet ports for airflow and an inlet port for temperature/humidity measurements (Fig. 1).

Test samples were tested in the chamber under the following test conditions:

- temperature 23°C + 2°C,
- relative humidity 45% ± 5%,
- air exchange rate $1 \times h^{-1}$,
- chamber loading factor 1 m^2/m^3,
- air velocity 0.2 m/s.

The chamber was placed in an air-conditioned room. Chamber air was sampled after 5, 24 and 48 hours.

Glass sorbent tubes were used to collect emissions. The tubes were packed with one layer of Tenax TA. Before use the tubes were conditioned at 260°C for 24 hours and exposed to the flow of helium at 15 ml min^{-1}.

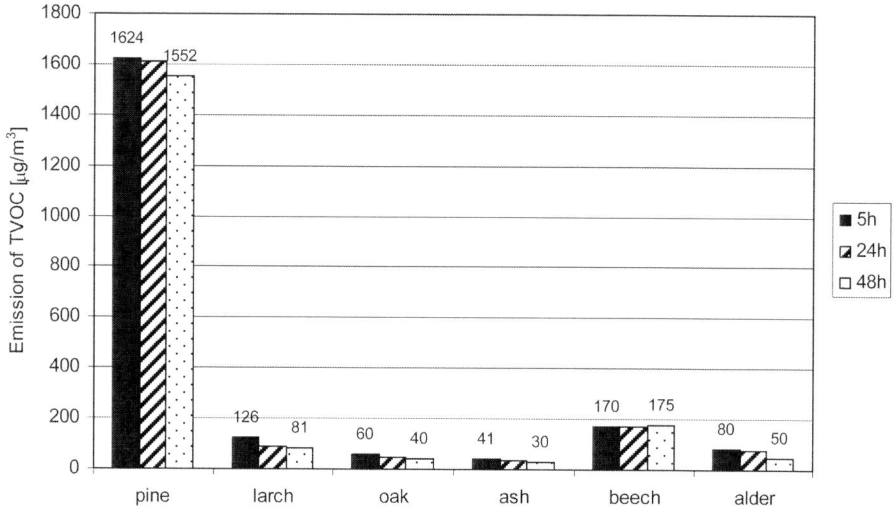

Figure 2. TVOC emission from tested wood species after 5, 24, 48 hours of exposure.

The sampling flow rate was set at 100 ml min^{-1}. Air (1 litre) sampled over 10 minutes was collected for analysis. After collection the substances adsorbed in the traps were thermally desorbed on-line into the chromatographic column.

The analysis was carried out using the GC/MS system (Thermoquest Finnigan Trace 2000) and coupled with thermal desorption unit (Master TD made by DANI Instruments), according to ISO 16000-6 (2004).

The analysis conditions were as follows:

- chromatograph: Trace 2000 THERMOQUEST/FINNIGAN,
- GC/MS interface temperature 220°C,
- detector MS 220°C,
- column: RTX-VMS (diameter 0.25 mm, phase thickness 1.4 µm, length 30 m),
- program temperature: 35°C (4 min) 5°C × min^{-1} → 140°C (0 min) 12° C × min^{-1} → 240°C (3 min),
- injector – thermal desorb unit – Master TD, 280°C.

The content of volatile organic compounds was expressed as TVOC sum determined in toluene equivalents based on calibration.

The target compounds were identified using the NITS MS library and an in-house database of the GC retention times for the compounds. Individual VOC and unidentifiable compounds emitted from wood and lacquered wood were quantified by the response factor of toluene and expressed as Toluene Equivalents (TE).

3 RESULTS AND DISCUSSION

The results of tests of TVOC emission from softwood and hardwood samples before lacquering and after 5, 24, and 48 h of exposure are given in Figure 2. Samples of pine wood emitted > ten-fold higher amounts of volatile organic compounds than the other tested wood species.

The sum of volatile organic compounds concentrations (TVOC) emitted from pine wood stayed within 1600 µg/m^3. In the analysed air mostly terpenes: α-, β-pinene, 3-carene, limonene were present (Tab. 1). The lowest amounts of volatile organic compounds were emitted by oak (60 – 40 µg/m^3) and ash (40 – 30 µg/m^3), and a little higher by beech (around 170 µg/m^3) and larch (120 – 80 µg/m^3).

The results of determinations of volatile organic compounds emission from coatings of water-based acrylic lacquer applied on tested hardwood and softwood species and on a glass panel are given in Figure 3 and in Table 2. After 5, 24 and 48 hours of exposure softwood samples with a

Table 1. Major organic compounds emitted from tested wood after 48 hours of exposure.

Chemical compound	CAS No.	Concentration in chamber air [$\mu g/m^3$]					
		Species of wood					
		pine	larch	oak	ash	beech	alder
Acetic acid	64-19-7			8	3	11	
4-Methylpentan-2-ol	72847-31-5				2		2
Toluene	108-88-3					11	3
1-Butanol	71-36-3	7		<1			
Butyl acetate	123-86-4		2			10	1
Hexanal	66-25-1	<1		1	1	9	4
p-Xylene	106-42-3	<1	2	<1		11	1
Styrene	100-42-5					9	
α-Pinene	80-56-8	481	30	<1	2	22	2
n-Propylbenzene	103-65-1					8	
1,2,4-Trimethylbenzene	95-63-6					9	
Benzaldehyde	100-52-7	20	5	7	6	18	7
Camphene	79-92-5	13					
β-Pinene	18172-67-3	349					
3-Carene	13466-78-9	473	1	<1	<1		4
Limonene	138-86-3	97	8	2		20	3
2-Ethyl-1-hexanol	104-76-7		5	3	1	4	3
n-Undecane	1120-21-4			1	2		
β-Phellandrene	99-83-2	40					
Acetophenone	98-86-2	7	5	5	2	6	6
2-Nonen-1-ol	31502-14-4	5	5	1			
Nonanal	124-19-6						4
n-Dodecane	112-40-3	6	3	1	2	<1	3
Naphthalene	91-20-3		4	1			<1

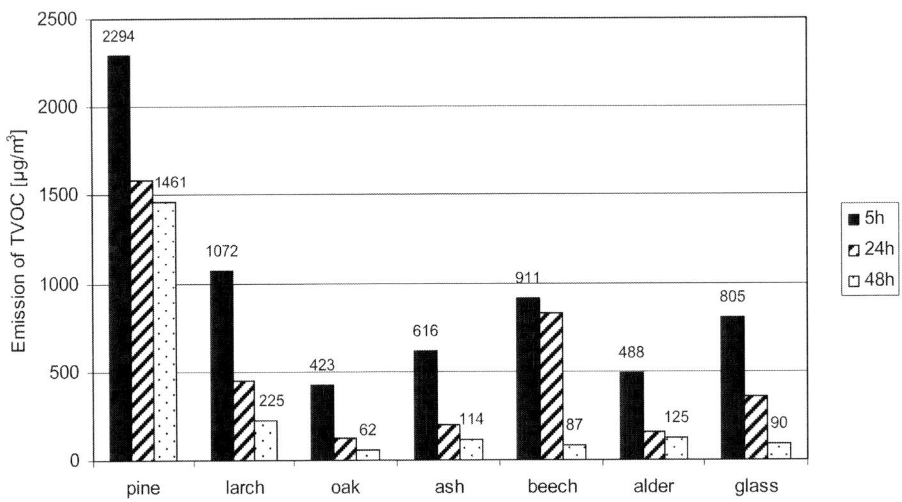

Figure 3. TVOC emission from acrylic lacquer coatings applied on various wood species and on a glass panel after 5, 24, 48 hours of exposure.

Table 2. Major organic compounds emitted from acrylic lacquer coatings applied to various wood species and to a glass panel after 48 hours of exposure.

		Concentration in chamber air [$\mu g/m^3$]						
		Species of wood						
Chemical compound	CAS No.	pine	larch	oak	ash	beech	alder	Glass
n-Hexane	110-54-3	<1	<1	4		1	8	
1-Butanol	71-36-3					4	2	
Toluene	108-88-3	<1	1	1	4	1	9	<1
Hexanal	66-25-1	17	<1	<1	<1	<1	6	
Butyl acetate	123-86-4	4	5	3	9	4	9	2
m-Xylene	108-38-3	<1	3	3	4	3	7	2
α-Pinene	80-56-8	551	42	4	6	5	12	1
n-Propylbenzene	103-65-1				4	3	4	
α-Pinene	18172-67-3	70	8					
3-Carene	13466-78-9	622	81		7	4		
1,2,4-Trimethylbenzene	95-63-6						20	
Limonene	138-86-3	29	9	3	6	4	8	2
1-Methoxy-2-propanol	107-98-2	81	65	20	68	49	34	74
n-Undecane	1120-21-4	5	4	7	4	5	6	3

water-based acrylic lacquer coating emitted to air mostly terpenes (>1200 $\mu g/m^3$), while emission of terpenes from lacquered pine wood (similar to the amounts emitted from unlacquered wood) was a few times higher than in the case of lacquered larch. In the case of hardwood species VOC concentrations observed in the air of the test chamber were significantly lower. After 48 hours of exposure TVOC for oak sample was only 62 $\mu g/m^3$. Slightly higher values were determined for lacquered alder – 125 $\mu g/m^3$ and ash – 114 $\mu g/m^3$. After 48 hours of exposure, softwood finished with acrylic lacquer coating emitted amounts of all volatile organic compounds TVOC (1461 $\mu g/m^3$ – pine, 225 $\mu g/m^3$ – larch) a few times higher than in the case of hardwood.

After 48 hours of exposure mostly terpenes (3-carene- 678 $\mu g/m^3$, α-pinene- 476 $\mu g/m^3$) were present in the air of the chamber in which the sample of pine wood finished with nitrocellulose lacquer was tested (Tab. 3). Additionally, in the composition of analysed air considerable amounts aromatic compounds (toluene- 198 $\mu g/m^3$, o-xylene- 272 $\mu g/m^3$) were identified. The highest values of concentrations were emitted by a lacquered beech wood sample – after 5 hours of exposure TVOC was approximately 10 000 $\mu g/m^3$, and after 48 hours it fell to around 2500 $\mu g/m^3$ (Fig. 4). Hardwood species finished with nitrocellulose lacquer emitted significant amounts of butyl acetate (beech ~600 $\mu g/m^3$, pine ~300 $\mu g/m^3$), whereas they did not emit significant amounts of terpenes and aldehydes. In that variant of the experiment the lacquered glass panel was characterised by the lowest volatile organic compounds emission, i.e. approx. 500 $\mu g/m^3$ after 5 hours of exposure and approx. 50 $\mu g/m^3$ after 48 hours of exposure (Fig. 4).

Chemo-hardened lacquers were another group of base and surface products. Most often these are two-component lacquers which release formaldehyde during hardening as a result of chemical reactions. Based on our analyses it was also observed that chemo-hardened lacquer emitted volatile organic compounds in amounts resulting in high concentrations of compounds identified in tested air (Tab. 4).

The concentration of volatile organic compounds determined in the chamber air for a pine wood sample finished with chemo-hardened lacquer ranged from around 4600 $\mu g/m^3$ after 5 hours of exposure to around 2600 $\mu g/m^3$ after 48 hours (Fig. 5). Although in the case of the other softwood species, i.e. larch, the composition of emitted compounds was similar in terms of quality, the total amount of volatile organic compounds in analysed air (TVOC) after 5 hours of exposure was the lowest compared to other samples and amounted to approx. 3000 $\mu g/m^3$; after two days it dropped to ~1100 $\mu g/m^3$ (Fig. 5).

In the case of oak, ash and alder samples finished with chemo-hardened lacquer even after only 5 hours of exposure the emission of volatile organic compounds was found to be much higher

Table 3. Major organic compounds emitted from nitrocellulose lacquer coatings applied on various wood species and on a glass panel after 48 hours of exposure.

		Concentration in chamber air [$\mu g/m^3$]						
		Species of wood						
Chemical compound	CAS No.	pine	larch	oak	ash	beech	alder	Glass
Ethyl acetate	141-78-6			3	5	37	6	
1-Methoxy-2-propanol	107-98-2	82	30	27	18	67	30	<1
Toluene	108-88-3	198	72	59	42	389	42	1
Acetic acid 2-methoxypropyl ester	108-65-6	50	13	14	5	166	6	
Hexanal	66-25-1	29	7	10	9		8	9
Butyl acetate	123-86-4	328	121	118	105	595	90	1
Ethylbenzene	100-41-4	60	15	9	4	126	7	
o-Xylene	95-47-6	272	71	45	23	425	35	2
n-Nonane	111-84-2					23		
2-Methyl-2-pentanol-4-one	123-42-2	46	6	7	6		7	<1
m-Xylene	108-38-3	125	28	20	8	244	14	
3-Ethoxy-1-propanol	111-35-3					20		32
α-Pinene	80-56-8	476	43		<1		<1	
2-Butoxyethanol	111-76-2		7	8	4		5	
1-ethyl-2-methylbenzene	611-14-3	23	<1	<1		72		
β-Pinene	18172-67-3	38	14					
3-Carene	13466-78-9	678	92					
Limonene	138-86-3	46	8		<1			
1-methyl-4-ethylbenzene	622-96-8	74	18	7	2			
1,2,3-Trimethylbenzene	526-73-8		<1	10	10	158	14	
n-Undecane	1120-21-4		4	13		22		

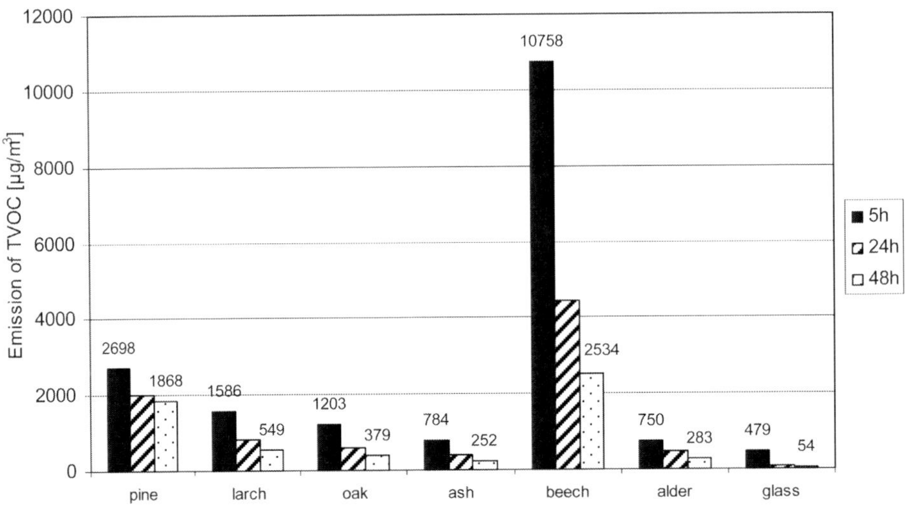

Figure 4. TVOC emission from nitrocellulose lacquer coatings applied on various wood species and on a glass panel after 5, 24, 48 hours of exposure.

compared to softwood samples. The value of emitted TVOC was approximately 6000 $\mu g/m^3$–7000 $\mu g/m^3$ and it was close to the value of TVOC determined for the glass panel. On the other hand, significant differences were observed after 48 hours of exposure when the sum of volatile organic compounds TVOC determined for lacquered oak sample dropped to the level of 2700 $\mu g/m^3$,

Table 4. Major organic compounds emitted from chemo-hardened lacquer coatings applied to various wood species and to a glass panel after 48 hours of exposure.

Chemical compound	CAS No.	Concentration in chamber air [$\mu g/m^3$]						
		Species of wood						
		pine	larch	oak	ash	beech	alder	Glass
3-methyl-4-penten-2-ol	4325-82-0	640	238	632	531	1363	269	
1-Butanol	71-36-3	<1	237	508	23		377	343
Toluene	108-88-3	23		85	24	325		
Acetic acid 2-methoxypropyl ester	108-65-6	17		32		319		
Butyl acetate	123-86-4	96	43	148	86	884	72	<1
Ethylbenzene	100-41-4	85	36	122	60	983	31	
o-Xylene	95-47-6	282	118	345	218	2101	110	9
m-Xylene	108-38-3	60		77	35	745	21	<1
α-Pinene	80-56-8	346	66					
2-Butoxyethanol	111-76-2	121	124	115	111	284	128	176
n-Propylbenzene	103-65-1	55	14	64	28	618	22	<1
1-Ethyl-2-methylbenzene	611-14-3	216	59	226	130	2269	101	19
1,2,4-Trimethylbenzene	95-63-6	65	24	58	25	545	26	2
3-Carene	13466-78-9	465						
1,2,3-Trimethylbenzene	526-73-8	26	68	158	150	1177	84	46
1,2,4,5-Tetramethylbenzene	95-93-2						14	26
n-Decane	124-18-5	75	30	68	31	60	42	10

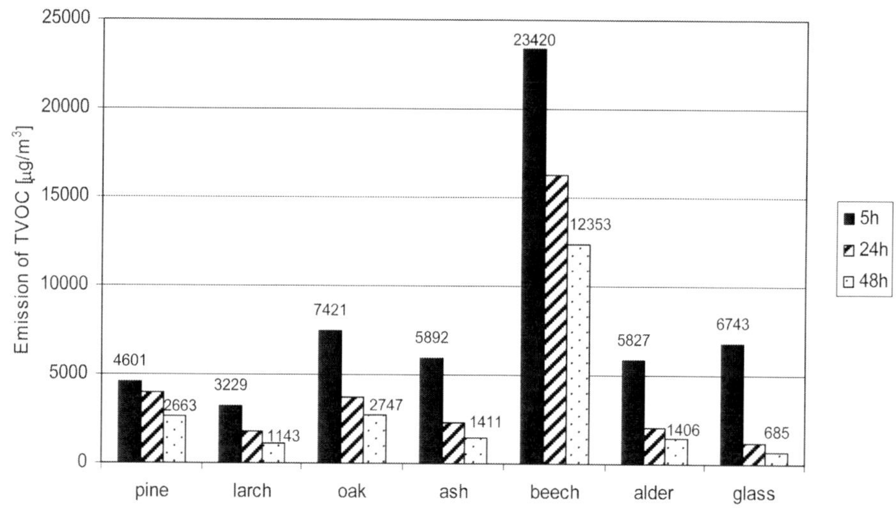

Figure 5. TVOC emission from chemo-hardened lacquer coatings applied on various wood species and on a glass panel after 5, 24, 48 hours of exposure.

and for ash and alder to 1400 $\mu g/m^3$. The result for the lacquered glass panel was ca. 700 $\mu g/m^3$ and was the lowest compared to TVOC values determined for other test samples (Fig. 5).

Similar to the case of nitrocellulose lacquer, chemo-hardened lacquer applied to a beech wood surface resulted in a coating emitting the highest amounts of volatile organic compounds – 23 000 $\mu g/m^3$ after 5 hours of exposure. After 48 hours of exposure TVOC concentration was 12 000 $\mu g/m^3$ (o-xylene ~2000 $\mu g/m^3$, 1-ethyl-2-methylbenzene ~2200 $\mu g/m^3$) and after that exposure time it was still the highest. Results of determinations of VOC emission from coatings

Table 5. Major organic compounds emitted from polyurethane lacquer coatings applied to various wood species and to a glass panel after 48 hours of exposure.

		Concentration in chamber air [$\mu g/m^3$]						
		Species of wood						
Chemical compound	CAS No.	pine	larch	oak	ash	beech	alder	Glass
Ethyl acetate	141-78-6					4	10	
1-Butanol	71-36-3	22	18	43	20	19	19	13
Toluene	108-88-3		6		6		29	
Butyl acetate	123-86-4	16	18	25	29	82	110	
Ethylbenzene	100-41-4	1076	543	751	884	838	724	2
o-Xylene	95-47-6	2402	1469	1823	2019	2542	2934	27
1-methoxy-2-propyl acetate	108-65-6	2050	1396	1480	1877	2701	1328	496
m-Xylene	108-38-3	499	256	316	420	610	790	
α-Pinene	80-56-8	88	17	37	1	<1		
1-ethyl-2-methylbenzene	611-14-3	9	11	35	3	11	2	3
2-Butoxyethanol	111-76-2		24		18			9
1,2,4-Trimethylbenzene	95-63-6	5	4	11	22	16	14	3
3-Carene	13466-78-9	82						
Limonene	138-86-3	2	<1	7		3		

of chemo-hardened lacquer applied to various materials, calculated keeping the division into basic groups of volatile organic compounds emitted, proved that, in the case of pine, considerable amounts of terpenes (3-carene- 465 $\mu g/m^3$, α-pinene- 346 $\mu g/m^3$) were emitted – Table 4.

Contrary to the above discussed variants with acrylic, nitrocellulose and chemo-hardened lacquers, VOC emission from polyurethane lacquer coating applied to pine wood did not yield high concentrations of terpenes in analysed air after 48 hours of exposure (Tab. 5).

The polyurethane lacquer coating applied on all analysed test samples mostly released esters (1-methoxy-2 propyl acetate) and aromatic compounds (m-xylene, o-xylene, ethylbenzene). The sum of volatile organic compounds TVOC emitted after 5, 24 and 48 hours of exposure was within the concentration range from approximately 4000 $\mu g/m^3$ to approximately 10 000 $\mu g/m^3$. The sum of volatile organic compounds emitted from lacquered pine was around 6000 $\mu g/m^3$ independent of the exposure time. In the case of larch, oak and alder a two-fold fall in concentration of volatile organic compounds was observed in results obtained after 48 hours of exposure of the sample in the chamber, while ash and beech, similarly to pine, did not demonstrate such dynamics of fall of TVOC concentration after another hour. The coating of polyurethane lacquer applied on the glass panel already in the first hours of exposure emitted a many-times lower amount of volatile organic compounds whose content in air dropped from 1400 $\mu g/m^3$ almost two-fold after 24 hours and three-fold after 48 hours (Fig. 6).

4 CONCLUSIONS

The emission of volatile organic compounds from systems consisting of lacquer and lacquered surface was tested for four types of lacquers applied to various species of softwood (pine and larch), hardwood (oak, ash, beech, and alder), and a glass panel.

After 48 hours of exposure the tests of chosen wood species intended for lacquering resulted in similar low values of TVOC emission in the case of larch, oak, ash and alder (30 $\mu g/m^3$ to 80 $\mu g/m^3$). In the case of beech, the results were twofold higher (175 $\mu g/m^3$), while the result for pine, which emits mainly terpenes, was high and equalled around 1500 $\mu g/m^3$.

As soon as after 5 hours of exposure the coating of water-based acrylic lacquer applied to individual wood species and to the glass panel was characterised by high VOC emission (mainly

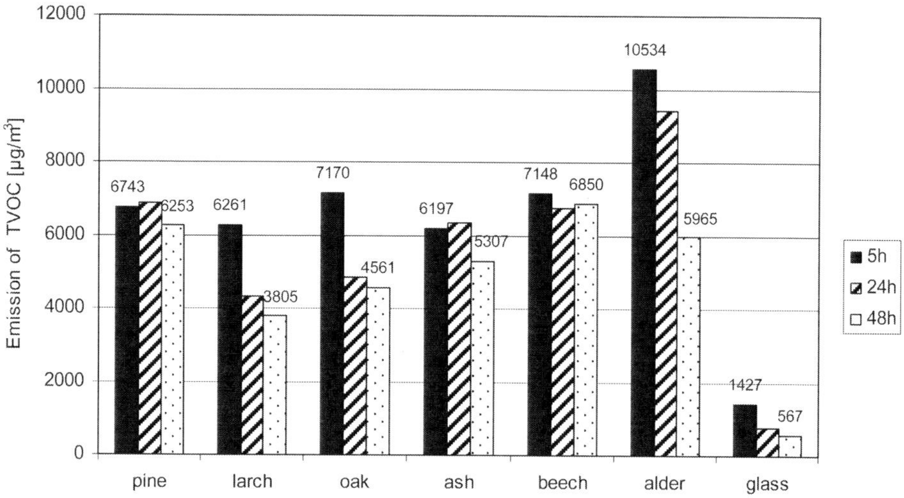

Figure 6. TVOC emission from polyurethane lacquer coatings applied on various wood species and on a glass panel after 5, 24, 48 hours of exposure.

derivatives of propanol), while after 48 hours the concentration of volatile organic compounds in the chamber air already reached values ranging from 60 $\mu g/m^3$ to ~1500 $\mu g/m^3$, thus they were similar to the results obtained in the tests of samples not finished with lacquer.

The highest TVOC values (after 48 hours of exposure) were obtained in the case of the beech sample finished with the following solvent lacquers: nitrocellulose (~2500 $\mu g/m^3$), chemo-hardened (~12 000 $\mu g/m^3$), and polyurethane (~7000 $\mu g/m^3$).

The obtained results indicate the crucial effect of characteristic anatomic features and chemical properties of beech wood on VOC emission from lacquer coatings (Wagenführ & Scheiber 1985).

Hence, the high emission of butyl acetate from lacquer coatings applied on beech wood may be connected with, for example, the relatively high content of acetic acid in the chemical composition of this wood species. However, confirmation of such hypotheses will require additional detailed tests and chemical analyses of wood samples in terms of the content of particular chemical compounds and possible reaction of these compounds with the compounds contained in lacquers used for finishing of surfaces of various wood species. It is characteristic that lacquer coatings applied on the glass panel, which is used in practice for assessment of lacquer hygienicity, demonstrated the lowest TVOC emission.

It shows, that the assessment of lacquer hygienicity, based on the tests of volatile organic compounds emission from coatings applied on chemically neutral, non-porous glass panel, cannot be a basis for unambiguous assessment of lacquers intended for finishing of elements made of wood or wood-based materials.

Hitherto no similar tests taking into consideration the effect of individual wood species on the emission of volatile organic compounds from lacquer coatings, have been carried out. The preliminary results we obtained indicate the significant influence of this factor on the presence of volatile organic compounds in the human environment, because in practice elements made of natural wood finished with lacquers are increasingly used in residential rooms. In the light of the present results, the materials used for equipping residential rooms need testing for volatile organics emission from lacquer coatings applied to surfaces of defined, widely used wood species of different properties.

REFERENCES

Englund, F. & Harderup, L.E. 1996. Indoor air VOC levels during the first year of a new three-story building with wooden frame. *Proc. Indoor Air, 7th Int. Conf. on Indoor Air Quality and Climate, Nagoya* 3: 47–52.

Englund, F., Larsen, A., Winther, L, Saarela, K. & Tirkkonen, T. 1996. Emissions of VOC from wooden floors, surface treated with oils and waxes. *Proc. Indoor Air, 7th Int. Conf. on Indoor Air Quality and Climate, Nagoya:* 89–94.
Gaca, P. & Dziewanowska-Pudliszak, A. 2005. The hazardous emission from material applied in furniture manufacture. *Proceedings, Indoor Air, Beijing:* 2294–2298.
Guo, H., Lee, S.C. & Kwok, N.H. 2002. Modeling of VOCs emissions from a varnish. *Proceedings: Indoor Air, Monterey CA*: 226–231.
Maupetit, F., Ramalho, O., Mienne, A., Kelly, P.O. & Yrieix C. 2002. Measurement of VOC emissions from wood remedial treatment: chamber tests and field experiments. *Proceedings: Indoor Air, Monterey CA*: 219–224.
Ohlmeyer, M. 2010. Emissions from Wood and Wood-based panels. European Symposium: Emissions of Regulated Dangerous Substances from Building Products, Hannover 16–17 March 2010.
Reiser R., Meile A., Hofer C. & Knutti R. 2002. Indoor Air Pollution by Volatile Organic Compounds (VOC) emitted from flooring material in a Technical University in Switzerland. *Proceedings: Indoor Air, Monterey CA*: 1004–1008.
Roffael E. 2006. Ursachen fur Emissionen aus Holz und Holzwerkstoffen. *Holz-Zentralblatt* 20: 630–631.
Uhde, E. 2010. Indoor air related evaluation of windows and doors. European Symposium: Emissions of Regulated Dangerous Substances from Building Products, Hannover 16–17 March 2010.
Wagenführ R. & Schreiber Ch. 1985. *Holzatlas.* VEB Fachbuch Leipzig.
Wilgusz R., Sitko E., Nikel G., Jarnuszkiewicz I. & Igielska B. 2002. The effect of temperature on the emission of formaldehyde and Volatile Organic Compounds (VOCs) from laminate flooring – case study. *Building and Environment* 37: 41–44.
Wilke, O. & Jann, O. 2002. VOC- and SVOC-emissions from adhesives, floor coverings and complete floor structures. *Proceedings: Indoor Air 2002, Monterey CA*: 962–967.

Management of Indoor Air Quality – Dudzińska (ed)
 ISBN 978-0-415-67266-5

Nail polishes and nail polish removers as sources of volatile organic compound emissions

E. Huzar & A. Wodnicka
Institute of Organic Chemical Technology, West Pomeranian University of Technology, Szczecin, Poland

ABSTRACT: Compounds which have an effect on indoor air quality may be emitted from different household products. Cosmetics are one of this group of products. The content of Volatile Organic Compounds (VOC) in nail polishes and nail polish removers available on the Polish market was determined using gas chromatography. In the nail polish removers analyzed, acetone, methanol, ethanol, isopropyl alcohol, ethyl and butyl acetates were present. In the analyzed nail polishes aliphatic esters, aliphatic alcohols and aromatic hydrocarbons were identified. The total VOCs emission intensity was determined using the gravimetric method. In order to evaluate indoor air quality, the concentrations of VOCs in a standard room were calculated and the inhalation exposure to VOCs was estimated.

Keywords: indoor air quality, volatile organic compounds, cosmetics, nail polish removers, nail polishes, inhalation exposure to VOCs

1 INTRODUCTION

Largely because of the increased incidence of Sick Building Syndrome (SBS) and other building-related illnesses, indoor air quality is of growing concern. Some studies have reported a correlation between symptoms of SBS and the total concentration of Volatile Organic Compounds (VOCs). Although indoor VOCs are recognized as potentially harmful and human carcinogens, a direct association between exposure to VOCs and SBS symptoms or cancer has not been fully established at typical indoor air concentrations (Guieysse et al. 2008). However, in the majority of cases the incidence of symptoms is higher at increased total VOCs concentration (Hutter et al. 2005, Wang et al. 2007) and several studies have correlated exposure to low concentrations of indoor pollutants with increased risks of cancer or other common symptoms e.g. headache, eye and respiratory system irritation (Guieysse et al. 2008, Wolkoff & Nielsen 2001).

Volatile Organic Compounds (VOCs) are defined as organic compounds with boiling points ranging from about 50°C to about 260°C (Wolkoff & Nielsen 2001). VOCs determined in indoor air belong to different groups of compounds. There are carbonyl compounds (aldehydes and ketones), aliphatic and aromatic hydrocarbons, terpenes, alcohols, esters, halogenated hydrocarbons, ethers and others (Wang et al. 2007, Wolkoff & Nielsen 2001, Zabiegala et al. 2003). In addition, many harmless VOCs can react with oxidants such as ozone, hydroxyl radicals and nitrogen oxides, producing highly reactive compounds that can be more harmful than their precursors (Guieysse et al. 2008, Kwon et al. 2007, Kwon & Jo 2007, Nazaroff & Weschler 2004).

Indoor VOCs concentrations are influenced by a large number of factors. The most important of these are characteristic of the sources of emission, air exchange rate and ventilation systems, activities of building occupants (e.g. cooking, smoking and photocopying), meteorological factors such as temperature and relative humidity, age of a building and its design. One essential factor is also the removal rate of VOCs, which depends on sorption, desorption and chemical reactions rates (Ayoko 2004).

Building materials have received considerable attention in connection with indoor air quality. However, for some compounds such as aromatics, the occupants of the building and their activities are more important to VOCs concentrations than ventilation systems and building materials (Ayoko 2004, Jurvelin 2003). These activities are very often associated with the use of liquid household products, which are recognized as sources of human exposure to certain chemicals (Kwon & Jo 2007, Kwon et al. 2007, Kwon et al. 2008, Zhu et al. 2001). The present study is focused on two groups of such products – nail polishes and nail polish removers.

Nail polish removers are used to remove any nail enamel at the beginning of the manicure procedure. According to literature, nail polish removers may contain acetone, ethyl acetate, butyl acetate or 2-butanone (Moossavi & Scher 2001). Investigations conducted by Tanaka-Kagawa et al. (2005) in Japan confirmed that these compounds are present at a frequency of 76–100% in indoor air; a comparison of indoor and outdoor concentrations of these compounds indicated an indoor source of emission.

Nail polishes are the second important group of products used during the manicure procedure. According to the literature solvent-extenders (solvents and diluents) make up about 70–80% of the liquid polish (Marcinkiewicz-Salmonowiczowa 1995, Moossavi & Scher 2001). They are a mixture of volatile and very volatile substances and allow the other components of the polish to remain in a liquid state. The most common solvents used in nail polishes are acetone, ethyl acetate, butyl acetate, toluene, isopropyl alcohol and ethylene glycol (Marcinkiewicz-Salmonowiczowa 1995, Moossavi & Scher 2001). As diluents ethyl and butyl alcohols, aliphatic and aromatic hydrocarbons are used (Marcinkiewicz-Salmonowiczowa 1995).

Taking into account the high content of volatile organic compounds in nail polishes and nail polish removers, emission from these products may be associated with health risks for building occupants. For that reason an attempt to determine the VOCs emitted from these products to the air and to estimate the concentration of VOCs in the air was undertaken. Calculation of inhalation exposure to VOCs associated with the use of nail polishes and polish removers requires considering different scenarios. The first takes into account the home manicure procedure and inhalation exposure to VOCs of a cosmetics user and remaining building occupants. The second scenario assumes the manicure procedure at a beauty salon and inhalation exposure to VOCs of salon employees and clients.

2 MATERIALS AND METHODS

2.1 *Samples*

Two groups of cosmetics used at manicure procedure were analyzed as potential sources of indoor air pollution. These were nail polish removers and nail polishes commonly available on the home market in Poland. A total of eight nail polish removers and ten nail polishes currently sold in Poland were analyzed. All samples of the nail polish removers were produced in Poland. The nail polishes were produced in different countries (Poland, France, Germany, Sweden, Spain, Turkey and the USA).

2.2 *Analysis of nail polish removers*

Studies of VOCs content in nail polish removers available on the home market were performed by gas chromatography. Investigations were conducted according to the method developed for the analysis of acetone in nail polish removers (Huzar & Wodnicka 2009). The applied method enables a selective determination of organic solvents in consumer products.

Samples of the product to be analyzed (5 μl nail polish remover) were injected through silicon membranes into glass tightly closed containers (1114 cm^3). The samples were placed in a thermostat chamber at 40°C and stabilized at room temperature for 30 minutes. In these conditions all the VOC evaporates to the gas phase. After equilibration, gas samples (1 cm^3) were drawn with a Hamilton gas-tight syringe and analyzed on a gas chromatograph. The qualitative and quantitative analyses of remover components were performed using a CHROM 5 gas chromatograph with a Flame-Ionization Detectors (FID). The chromatograph was provided with two steel chromatographic columns. Column I was 2 meters long and of 3 mm inside diameter packed with 8% SE-30 + 0.21% Carbowax 20M on Chromosorb W NAW (60–80 mesh). Column II was 1 meter long and of 3 mm inside diameter packed with Chromosorb 102 (80–100 mesh). The following temperatures were applied: column I 50°C, column II 100°C, injectors 120°C, detectors 150°C. Nitrogen (40 $cm^3 \cdot min^{-1}$) was used as the carrier gas.

Identification of the ingredients of removers was performed on both columns using acetone, methyl, ethyl and isopropyl alcohols, ethyl acetate, n-butyl acetate as standards. All were of analytical grade.

Quantitative analyses of aliphatic esters were performed on column I. On the column I peaks of acetone and other volatile compounds such as ethyl and isopropyl alcohols were not separated very well. Analyses of acetone, methyl, ethyl and isopropyl alcohols were performed on column II. The apparatus was calibrated using the standard gas mixtures prepared in the same containers and conditions in which the tested samples were prepared. Two standard mixtures were used for calibration. The first mixture contained methyl, ethyl and isopropyl alcohols and acetone (1:1:2:2). In the second, ethyl and butyl acetates (1:2) were used with acetone as a solvent. After evaporation and stabilization, gas mixtures of 0.4; 0.8; 1.2 and 1.6 cm^3 volume were drawn with a Hamilton gas-tight syringe and analyzed on a gas chromatograph. The obtained calibration curves were rectilinear with the correlation coefficients in the range of 0.987–0.996.

The content of ingredients (*CP*) in the examined products was calculated from the following equation:

$$CP = \frac{m \cdot V_C}{V_S \cdot V_R \cdot d} \quad (\mathrm{mg \cdot g^{-1}}) \tag{1}$$

where m = mass of the substance calculated from its calibration curve (mg); V_C = volume of the container (cm^3); V_S = volume of gas sample injected to the chromatographic column (cm^3), V_R = volume of nail polish remover injected into container (cm^3); d = density of nail polish remover determined for 10 cm^3 of product ($g \cdot cm^{-3}$).

The total emission intensity of VOCs was determined using the gravimetric method. To this end, nail polish removers were applied on cotton pads and the polish removing procedure was simulated. The weight loss of removers from cotton pads was observed. The intensity of VOCs emission was determined as a slope of relationships between weight loss and the time in the range of 1–20 minutes.

2.3 *Analysis of nail polishes*

Analysis of volatile organic compounds evolved from nail polishes included three experimental steps. First, the total VOCs content was determined using the gravimetric method. To this purpose, the nail polishes were spread on microscope slides (76 mm × 26 mm) and the weight loss due to evaporation of volatile compounds was observed. After equilibrium was reached, the percentage content of VOCs in samples was calculated and the intensity of VOCs emission was determined from the slope of the relationship between weight loss and time in the range of 1–5 minutes.

The identification of VOCs emitted from nail polishes to the air was the second step of the studies. It was performed using GC-MS. An Agilent 6890 N gas chromatograph with a 5973 N mass selective detector, equipped with a 7683 series injector was applied. In order to prepare samples for the GC-MS analyses, volatile substances liberated from nail polishes were adsorbed on active carbon (Merck 35–50 mesh) using the aspiration method. Extraction from the active carbon was performed using two different solvents: acetone (for analysis of the less-volatile compounds) and n-butyl acetate (for very volatile compounds). A HP-5MS capillary column with 5% phenyl-methylpolysiloxane, 30 m × 0.25 mm I.D. and 0.25 μm film thickness was used with the following temperature program: 50°C (hold 5 min), 10°C min^{-1} to 280°C (hold 2 min). Helium was used as carrier gas at a constant flow rate of 1.2 cm^3 min^{-1}. The injection port temperature was 250°C, 2 μl of solution was injected via split (1:10) injection. The mass selective detector conditions were as follows: electron impact ionization, full scan mode (10–300 m/z), MSD transfer line temperature: 280°C, MS quad: 150°C; MS source: 230°C. The mass selective detector was turned off during the elution time of solvents (2.10–9.00 min. for n-butyl acetate and 0–1.80 min. for acetone). Compounds were identified using a NIST 02 mass spectral library. A spectral search quality of 90% was set as a criterion for compounds selections and identification was confirmed by measuring of the retention times of suitable standards.

The last step of the investigations was quantitative analyses of VOCs emitted from analyzed samples of nail polishes. The samples of nail polishes (1–2 drops) were placed in 1.75 cm^3 glass

Table 1. Density, volume and mass detector response factors of analyzed substances.

Group of analytes	Compound	Density ($g \cdot cm^{-3}$)	Volume detector response factor* f_V	Mass detector response factor f_G
Aliphatic alcohols	Ethyl alcohol	0.7894	1.819	1.436
	Isopropyl alcohol	0.7864	1.311	1.031
	n-Butyl alcohol	0.8098	0.997	0.807
Aliphatic esters	Ethyl acetate	0.9010	1.488	1.341
	Isopropyl acetate	0.8718	1.182	1.030
	n-Propyl acetate	0.8867	1.182	1.048
	Isobutyl acetate	0.8712	1.000	0.871
	n-Butyl acetate	0.8813	1.000	0.881
Aromatic hydrocarbons	Toluene	0.8670	0.556	0.482
	Ethylbenzene	0.8669	0.538	0.466
	p-Xylene	0.8610	0.538	0.463
	o-Xylene	0.8801	0.538	0.473

* the relative detector response factors of isomers was taken as identical.

vials, tightly closed with the screw caps with silicon/PTFE membranes and equilibrated to room temperature for 1 hour. Due to the small gas phase volume, head space sampling was performed by manual pressurization with a Hamilton gas-tight syringe. During the initial step, the syringe needle was inserted into the vial and the vial content was pressurized with 0.5 cm^3 of clean air. After equilibration the gas sample was drawn and 0.3 cm^3 of sample was analyzed using a CHROM 5 gas chromatograph. Analyses were performed on column I under the same conditions as applied during analyses of nail polish removers.

The quantitative analyses were performed by an internal normalization method. To determine the volume relative detector response factors two standard liquid mixtures containing equal volumes of substances were used. The first standard mixture (I) was prepared using ethanol, ethyl acetate, isopropyl acetate, toluene and n-butyl acetate. Standard mixture II contained isopropyl alcohol, n-butyl alcohol, n-butyl acetate and p-xylene. Volume response factors were calculated as the ratio of the peak area of the determined substance to the peak area of n-butyl acetate (the volume response factor of n-butyl acetate was taken as 1). Mass relative detector response factors were calculated as follows:

$$f_{Gi} = f_{Vi} \cdot d_i \quad (2)$$

where f_{Gi} = the mass relative detector response factor of the substance i; f_{Vi} = the volume relative detector response factor of the substance i; and d_i = density of the substance i.

The detector response factors of analyzed compounds are presented in Table 1.

3 RESULTS AND DISCUSSION

3.1 *VOCs content in nail polish removers*

In analyzed samples acetone occurred with the highest frequency (75%). Ethyl acetate was observed in three samples. This corresponds to 37.5% frequency of occurrence. Although three producers declared the content of butyl acetate in their products, this compound was not identified in any sample. In analyzed samples the presence of aliphatic alcohols (methyl, ethyl and isopropyl alcohols) was also confirmed. Their frequencies of occurrence varied from 25 to 37.5%.

Previous work discusses the VOCs emitted from nail polish removers. Zhu et al. (2001) reported the presence of ethyl acetate, toluene and 2-butoxyethanol in nail polish removers available on the Canadian market. Investigations conducted in Korea show that from this group of products acetone, 2-butoxyethanol, chlorobenzene, 1,4-dioxane, ethanol, hexane, limonene, tetrachloroethylene,

Table 2. VOCs content in analyzed nail polish removers.

Product	Density of remover ($g \cdot cm^{-3}$)	Content ($mg \cdot g^{-1}$)				
		MA	EA	iPA	Acetone	EAc
R1	0.861	–	–	–	738.77	–
R2	0.889	–	191.02	–	–	681.66
R3	0.848	–	416.48	–	–	573.21
R4	0.801	708.50	–	–	221.83	–
R5	0.831	–	–	359.01	596.54	–
R6	0.800	732.62	–	46.08	145.58	–
R7	0.789	–	–	438.80	533.67	–
R8	0.851	–	–	–	14.58	453.02
Frequency (%)		25	25	37.5	75	37.5

MA: methyl alcohol, EA: ethyl alcohol, iPA: isopropyl alcohol, EAc: ethyl acetate.

phenol, 1-propanol, trichloroethylene and o-xylene may be emitted (Kwon & Jo 2007, Kwon et al. 2008). In another article, the same authors reported acetone, cyclohexane, butyl acetate and 2,2,4-trimethyl-1,3-dioxolane emission from nail polish removers (Kwon et al. 2007). These results and our investigations indicate that the chemical components emitted from nail polish removers may vary in different products produced in different countries.

VOCs content in analyzed products was calculated from Equation 1. The results are presented in Table 2. Coefficients of variation for these results did not exceed 5%. The results indicate that this group of products might be a substantial source of VOCs emission. The total content of VOCs in the majority of samples was very high and varied from 87 to 99% (w/w). Only in the samples R1 and R7 was the VOCs content lower (74 and 58% (w/w)), respectively.

3.2 *Content of VOCs in nail polishes*

The total VOCs content in the analyzed nail polishes determined using the gravimetric method is presented in Figure 1. It was in the range 65–73%. The average value of VOCs content in analyzed samples was $69.91 \pm 1.74\%$ (w/w) at level of significance 0.05. The coefficient of variation was 4%.

The number of chemicals emitted from the nail polishes analyzed ranged from 5 to 9. Four compounds (ethyl alcohol, isopropyl alcohol, ethyl acetate and n-butyl acetate) were present in all samples. In two nail polishes (S3 and S8) aromatic hydrocarbons were identified. Toluene was found in sample S8 and ethylbenzene, p-xylene, o-xylene were present in sample S3. Moreover, the analyzed samples contained n-butyl alcohol, isopropyl acetate, n-propyl acetate and isobutyl acetate. The chemical composition of volatiles emitted from the nail polishes determined in the current study was different from that reported in the other studies.

Kwon et al. (2007) reported investigations of volatiles emitted from nail polishes. They identified benzene, toluene, ethyl alcohol, isopropyl alcohol, ethyl acetate and limonene. In other studies acetone, 2-butoxyethanol, ethanol, ethylbenzene, o-xylene, phenol and limonene were determined (Kwon & Jo 2007). In our investigations the presence of acetone, benzene, limonene, 2-butoxyethanol and phenol in nail polishes was not confirmed.

Quantitative analyses of substances emitted from nail polishes were conducted by using the internal normalization method. The percentage average mass contents of VOCs in the gas phase are presented in Table 3. Coefficients of variation for all results were below 5%.

Ethyl acetate was emitted from the nail polishes samples in the highest amount. The fraction of this compound in VOCs for the majority of samples was in the range 61–79%. In only two cases it was smaller – almost 40% for sample S8 and above 11% for sample S3. The second most important substance emitted from all samples to the gas phase was n-butyl acetate. Its content in VOCs varied from 9 to 21%. Emission of isopropyl alcohol was also at a high level (10–16%). The fractions of ethyl alcohol, n-butyl alcohol and isopropyl acetate in VOCs did not usually exceed 2.5%. Only the ethyl alcohol content in the S3 sample was higher (38%). Isobutyl and n-propyl

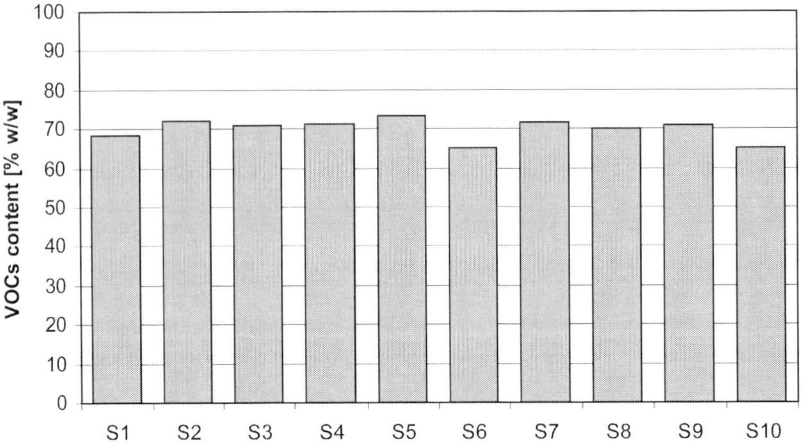

Figure 1. VOCs mass percentage content in analyzed nail polish samples.

Table 3. Mass percentage content of VOCs in the gas phase of analyzed nail polish samples.

	Mass percentage content (% w/w)											
	Aliphatic alcohols			Aliphatic esters					Aromatic hydrocarbons			
Product	EA	iPA	n-BA	EAc	iPAc	n-PAc	iBAc	n-BAc	T	EB	p-X	o-X
S1	1.24	10.13	0.55	79.02	0.13	–	–	8.93	–	–	–	–
S2	2.50	12.70	2.31	60.56	1.29	–	–	20.64	–	–	–	–
S3	38.24	10.32	0.30	11.79	1.97	–	–	19.37	–	3.39	12.48	2.14
S4	1.31	11.13	1.11	68.58	–	0.54	4.04	13.30	–	–	–	–
S5	1.26	12.87	0.48	63.91	0.39	–	4.00	17.09	–	–	–	–
S6	1.57	13.12	0.93	63.91	0.62	–	–	19.85	–	–	–	–
S7	0.81	12.49	–	65.34	–	11.44	–	9.91	–	–	–	–
S8	0.52	15.75	0.39	39.80	0.95	–	–	15.93	26.66	–	–	–
S9	0.76	15.96	0.27	68.83	0.09	0.17	–	13.91	–	–	–	–
S10	1.07	12.07	0.43	67.65	0.17	–	–	18.60	–	–	–	–
Frequency (%)	100	100	90	100	80	30	20	100	10	10	10	10

EA: ethyl alcohol, iPA: isopropyl alcohol, n-BA: n-butyl alcohol, EAc: ethyl acetate, iPAc: isopropyl acetate, n-PAc: n-propyl acetate, iBAc: isobutyl acetate, n-BAc: n-butyl acetate, T: toluene, EB: ethylbenzene, p-X: p-xylene, o-X: o-xylene.

acetates were present only in a few samples. Their content in gas phases was insignificant. The sample S8 requires special attention. In this sample the presence of toluene was observed. Toluene is a substance which is recognized as a priority indoor air pollutant by several organizations such as the WHO, the OQAI (French Indoor Air Quality Observatory) and the US EPA (Guieysse et al. 2008). In the substances emitted from sample S8, toluene made up 26.66%. This corresponds to 18.72% of toluene in the raw product. Aromatic hydrocarbons were also present in sample S3. From this sample ethylbenzene, p-xylene and o-xylene were emitted. The total content in VOCs was 18.01%.

The mass content of particular substances in raw products was calculated on the basis of the total VOCs content in the nail polishes and the mass percentage content of substances evolved from the analyzed samples. The results are collected in Table 4.

Table 4. VOCs content in the raw nail polishes.

Product	Content of VOCs in nail polishes ($mg \cdot g^{-1}$)											
	Aliphatic alcohols			Aliphatic esters					Aromatic hydrocarbons			
	EA	iPA	n-BA	EAc	iPAc	n-PAc	iBAc	n-BAc	T	EB	p-X	o-X
S1	8.50	69.41	3.77	541.45	0.89	–	–	61.19	–	–	–	–
S2	18.06	91.76	16.69	437.55	9.32	–	–	149.12	–	–	–	–
S3	271.62	73.30	2.13	83.74	13.99	–	–	137.59	–	24.08	88.65	15.20
S4	9.35	79.41	7.92	489.32	–	3.85	28.83	94.90	–	–	–	–
S5	9.22	94.21	3.51	467.82	2.85	–	29.28	125.10	–	–	–	–
S6	10.21	85.31	6.05	415.54	4.03	–	–	129.07	–	–	–	–
S7	5.80	89.39	–	467.64	–	81.88	–	70.93	–	–	–	–
S8	3.65	110.58	2.74	279.44	6.67	–	–	111.85	187.18	–	–	–
S9	5.38	112.90	1.91	486.90	0.64	1.20	–	98.40	–	–	–	–
S10	6.98	78.70	2.80	441.08	1.11	–	–	121.27	–	–	–	–

EA: ethyl alcohol, iPA: isopropyl alcohol, n-BA: n-butyl alcohol, EAc: ethyl acetate, iPAc: isopropyl acetate, n-PAc: n-propyl acetate, iBAc: isobutyl acetate, n-BAc: n-butyl acetate, T: toluene, EB: ethylbenzene, p-X: p-xylene, o-X: o-xylene.

3.3 *Estimation of volatile organic compounds concentration in the air*

Assuming that the emission of VOCs from the liquid products follows first order decay and that there is no losses by chemical reactions, adsorption and desorption processes of emitted compounds on the inner surface, the concentration of the volatiles in a room can be determined by the following equation (adapted from Zhu et al. 2001):

$$C = \frac{EF \cdot M}{V(N-k)} \cdot (e^{-kt} - e^{-Nt}) \tag{3}$$

where: C = concentration in a room ($mg \cdot m^{-3}$); EF = emission factor ($mg \cdot g^{-1} \cdot min^{-1}$); M = mass of product (g); V = volume of the room (m^3); N = exchange rate of air in the room ($min.^{-1}$); k = emission decay constant ($min.^{-1}$); t = elapsed time (min.).

In our investigations emission factors were determined during simulated nail polish removing and nail polish application procedures. Mass changes of products were observed. As an example, the mass of remover R1 emitted to the air in the ratio to the mass of product used in the experiment (M) in the time is presented in Figure 2.

The emission factor (EF) was determined as the slope of the linear relationship between the relative emission and the time in the range 1–20 minutes. The intercept of the relationship shown corresponds to an initial emission of remover ingredients during the removal procedure (EF_0). The intensity of VOCs emission from nail polishes was determined on the basis of the weight loss in the range of time 1–5 minutes. Higher intensity VOCs emissions occurred from polishes than from polish removers.

The determined values of emission factors for all samples are collected in Tables 5 and 6.

Taking into account determined VOCs emission factors (EF_0 and EF), the concentration of VOCs in a room can be expressed as:

$$C = \left(\frac{EF_0 \cdot M}{V} + \frac{EF \cdot M}{V(N-k)} \right) \cdot (e^{-kt} - e^{-Nt}) \tag{4}$$

The average concentration-time profiles of VOCs in the standard rooms per task (removing of the nail polish and nail polish application) were reconstructed using average values of the emission

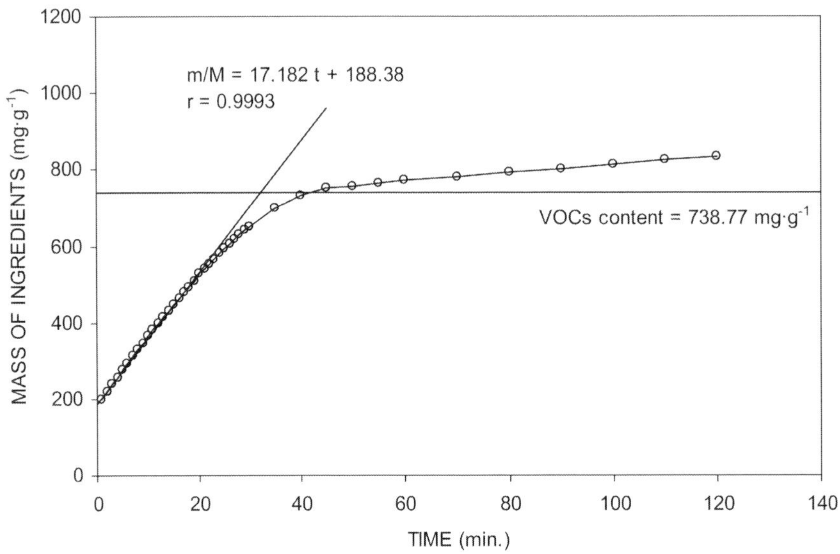

Figure 2. Mass changes of sample R1 components emitted to the air as a function of time.

Table 5. VOCs emission from the samples of nail polish removers.

Product	Total VOCs content, $(mg \cdot g^{-1})$	EF_0, $(mg \cdot g^{-1})$ (intercept)	EF, $(mg \cdot g^{-1} \cdot min.^{-1})$ (slope)	Linear regression coefficient, r
R1	738.77	188.38	17.182	0.9993
R2	872.68	370.68	24.168	0.9957
R3	989.69	83.39	15.736	0.9972
R4	930.33	253.21	17.666	0.9969
R5	955.55	142.08	18.863	0.9963
R6	878.20	180.26	13.064	0.9976
R7	579.75	124.28	13.853	0.9993
R8	906.40	232.55	27.046	0.9986
Average	856.42	196.85	18.45	

Table 6. VOCs emission from samples of nail polishes.

Product	Total VOCs content, $(mg \cdot g^{-1})$	EF_0, $(mg \cdot g^{-1})$ (intercept)	EF, $(mg \cdot g^{-1} \cdot min.^{-1})$ (slope)	Linear regression coefficient, r
S1	685.18	236.93	76.748	0.9699
S2	722.47	184.46	59.653	0.9930
S3	710.34	114.89	49.434	0.9899
S4	713.51	193.30	89.216	0.9720
S5	731.96	193.47	74.593	0.9763
S6	650.22	163.08	49.073	0.9843
S7	715.67	68.16	75.206	0.9799
S8	702.12	200.65	63.043	0.9899
S9	707.42	154.84	81.121	0.9689
S10	651.96	98.30	51.142	0.9920
Average	699.09	160.81	66.923	

factors (EF_0 and EF) and decay constant (k) for a given mass of product (M) and room conditions (V and N) according to Equation 4. The decay constant values were calculated from the following equation (Zhu et al. 2001):

$$k = \frac{EF \cdot M}{W} \tag{5}$$

where: W = the total evaporable amount of VOCs in the source (mg).

A mass of 3 g of nail polish removers and 0.4 g of nail polishes per task to calculations was taken: W values were estimated based on the total VOCs content in the analyzed products. The average amounts of these products used during the manicure procedure were determined on the basis of simulations. In addition, an average air exchange rate of 0.0083 min.$^{-1}$ (0.5 h^{-1}) and three different room volumes were used to calculate VOCs concentrations (C) in the air. According to standards and the literature the room volumes of 17.4 m^3 (Jensen & Knudsen 2006; Wolkoff & Nielsen 1996), 30.0 m^3 (Oppl et al. 2010) and 52.2 m^3 (Jensen & Knudsen 2006) were taken.

The concentration-time profiles of VOCs emitted from the analyzed products during a home manicure procedure are presented in Figure 3. These profiles correspond to a scenario in which the first step is removal of nail polishes (10 min.) and application of nail polishes is the second step.

The other scenario assumes the manicure procedure at a beauty salon. In this case the VOCs concentration in the ambient air follows from the cumulative emission associated with repeated manicure procedures. It was assumed that the individual procedures begin with an interval of one hour. The exemplary changes of VOCs concentration in the work time (8 hrs) are presented in Figure 4 and estimated 1-h concentrations of VOCs at the beauty salon are presented in Figure 5.

The results of estimation of VOCs concentration (Table 7) indicate that nail polish removers and nail polishes may be substantial sources of VOCs emission.

The studies of Kjærgaard et al. (1991) show the air quality to be unpleasant at concentrations of VOCs mixture above 8 mg · m^{-3}. For an exposure period of 50 min. at concentrations of 8 mg · m^{-3} or higher, irritation of the mucous membranes also was statistically significant. All results of our calculations of average 1-h VOCs concentration were above this value. Additionally, according to the scenario of manicure in a beauty salon, the values of VOCs concentration were high and varied from 68 to 203 mg · m^{-3}. Mølhave (1990) suggested a toxic range of VOCs concentration above 25 mg · m^{-3}, thus, irrespective of the room volume, beauty salons will require additional ventilation or VOCs removal systems to maintain acceptable indoor air quality. The estimated average 1-h concentrations for the home manicure scenario were certainly lower and in two cases (room volumes of 30.0 and 52.2 m^3) were in the discomfort range of VOCs concentration, which is defined as 3–25 mg · m^{-3} (Mølhave 1990). Home manicuring in small spaces such as bathrooms (17.4 m^3) results in exposure to high VOCs concentrations. The estimated average 1-h VOCs concentration at 17.4 m^3 room volume was in the toxic range (34 mg · m^{-3}). Opening a window to air the room seems to be the best way to improve the indoor air quality at home, but unfortunately in Poland the bathrooms of flats rarely have windows. To achieve the concentration in the multifactorial exposure range (0.2–3 mg · m^{-3}) or comfort range (below 0.2 mg · m^{-3}) (Mølhave 1990) additional ventilation, to enlarge the air exchange rate, would have to be installed.

3.4 *Estimation of exposure to VOCs associated with using of nail polish removers*

Due to the short task durations, the estimated 50th percentile 1-h concentrations of VOCs were selected for estimating of inhalation exposure. Additionally it was assumed that users of the products and other occupants of the room would be exposed during the duration of the whole manicure procedure (c.a. 40 minutes). The daily average exposure as a function of body weight was calculated from Equation 6.

$$DAEP = \frac{C_{50th} \cdot ADT \cdot AFT \cdot IR}{b.w.} \tag{6}$$

where: $DAEP$ = daily average exposure by inhalation per product (mg · (kg b.w.)$^{-1}$ · day^{-1}); C_{50th} = 50th percentile 1-h concentration of VOCs in the standard room (mg · m^{-3}); ADT = average

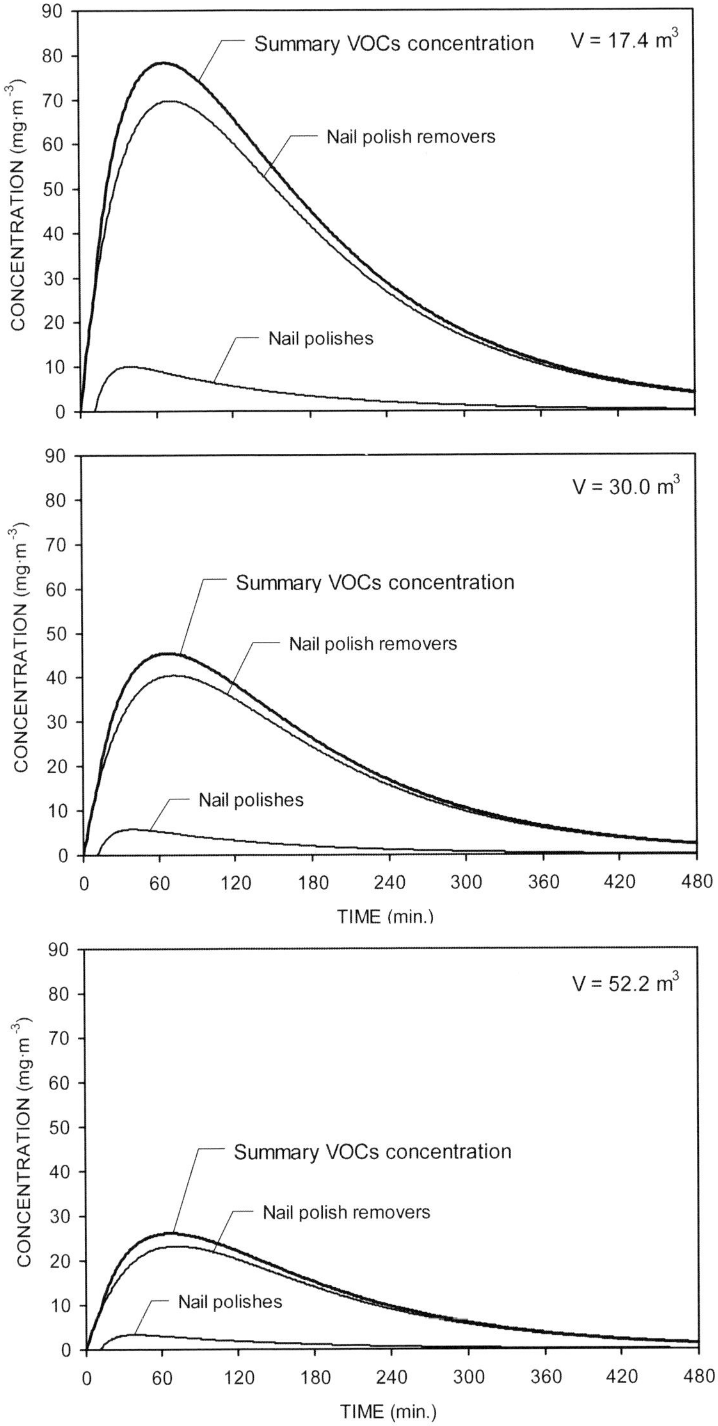

Figure 3. Concentration-time profiles of VOCs emitted during a single manicure procedure from 3 g of nail polish removers and 0.4 g of nail polishes (air exchange rate $0.5\,h^{-1}$, volume of the room: 17.4, 30.0 and $52.2\,m^3$).

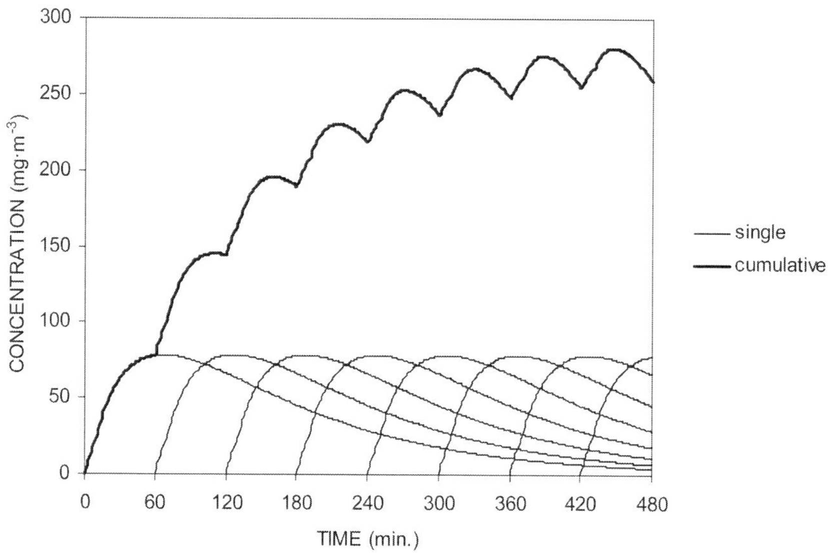

Figure 4. The cumulative VOCs concentration at a beauty salon resulting from repeated manicure procedures in a room of volume 17.4 m^3.

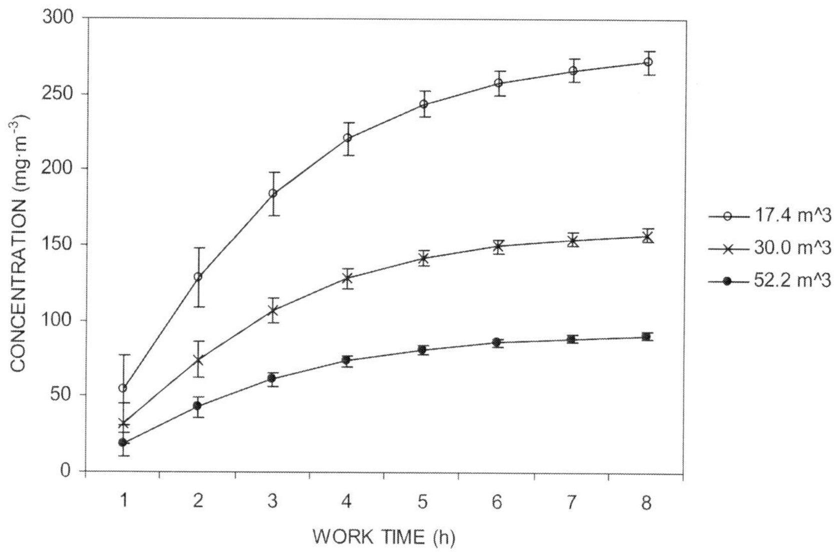

Figure 5. Estimated 1-h VOCs concentrations associated with the manicure procedure at a beauty salon.

duration per task (h/task); AFT = average frequency of the task (events/day); IR = inhalation rate ($m^3 \cdot h^{-1}$); $b.w.$ = body weight (kg).

For calculations it was assumed that 3 g of nail polish removers and 0.4 g of nail polishes are applied and the exposure time per event is 40 minutes (0.667 h/task) for a client and 1h for a salon staff member. The frequency of performing the task for the home scenario was taken as once every three days (0.33 events/day). The second scenario assumed eight manicures per beautician per day and one task per week per client (0.143 events/day). The estimated values of daily exposure to VOCs associated with the use of nail polish removers and nail polishes are collected in Table 8. Estimated average daily intake (DAEP) of volatile compounds for adults at home was 20–60 $\mu g \cdot (kg\ b.w.)^{-1} \cdot day^{-1}$. For a 6-year-old child this value was almost four-times

Table 7. Average 1-h VOCs concentrations in rooms used for different manicure procedures.

Scenario	Volume (m^3)	Concentration ($mg \cdot m^{-3}$) AM	SD	Min.	Max.	50th perc.	90th perc.
Home	17.4	33.98	25.26	5.31	74.12	29.84	61.11
	30.0	19.71	14.65	3.08	42.99	17.31	35.45
	52.2	11.33	8.42	1.77	24.71	9.95	20.37
Beauty salon	17.4	203.38	77.27	54.28	271.86	232.09	268.14
	30.0	117.96	44.82	31.48	157.68	134.61	155.52
	52.2	67.79	25.76	18.09	90.62	77.36	89.38

AM: arithmetic mean, SD: standard deviation, Min.: minimum value, Max.: maximum value, 50th perc.: 50th percentile concentration, 90th perc.: 90th percentile concentration.

Table 8. Estimated inhalation exposure to VOCs associated with use of nail polishes and nail polish removers.

Scenario	Volume (m^3)	Estimates of daily intake by inhalation, DAEP ($mg \cdot (kg\ b.w.)^{-1} \cdot day^{-1}$) Adult male[a]	Adult female[b]	Child, age 6 years[c]
Home	17.4	0.068	0.051	0.235
	30.0	0.039	0.029	0.136
	52.2	0.023	0.017	0.078
Beauty salon (client)	17.4	0.266	0.169	–
	30.0	0.131	0.098	–
	52.2	0.075	0.056	–
Beauty salon (staff)	17.4	19.019	14.195	–
	30.0	11.031	8.233	–
	52.2	6.339	4.731	–

[a] assuming an inhalation rate by light activity level 0.8 $m^3 \cdot h^{-1}$ and body weight 78.1 kg (U.S. EPA 1997)
[b] assuming an inhalation rate by light activity level 0.5 $m^3 \cdot h^{-1}$ and body weight 65.4 kg (U.S. EPA 1997)
[c] assuming an inhalation rate by light activity level 0.8 $m^3 \cdot h^{-1}$ and body weight 22.6 kg (U.S. EPA 1997)

higher. This results, in the first instance, from the low body weight. The estimated DAEP of VOCs at a beauty salon for clients was 66–218 $\mu g \cdot (kg\ b.w.)^{-1} \cdot day^{-1}$. The value of DAEP for an employee is a result of exposure time, which depends on the working hours in the establishment. In our calculations 8-hrs was assumed, so DAEP was in the range of 5.5–16.6 $mg \cdot (kg\ b.w.)^{-1} \cdot day^{-1}$. The exposure to VOCs for children at a beauty salon was not considered.

Uncertainties inherent in determining average daily intakes by inhalation associated with the use of nail polish removers and nail polishes are recognized.

There is only sparse information about exposure-effect relationships for a range of VOC mixtures in the literature, so that evaluation of our results is difficult. However, values of DAEP for the home manicure scenario were lower in comparison to the reference doses (RfD) of a particular volatile organic compounds (e.g. toluene – 223 $\mu g \cdot (kg\ b.w.)^{-1} \cdot day^{-1}$, xylenes – 200 $\mu g \cdot (kg\ b.w.)^{-1} \cdot day^{-1}$). Bearing in mind that VOCs emitted from nail polishes and nail polish removers are complex mixtures, even if the reference doses for mixtures existed, they would not be exceeded. In this case automatic measurements of TVOC (Total VOCs) would be useful to evaluate indoor air quality.

Unfortunately, automatic measurements of TVOC would not give complete information about potential exposure effects to VOCs at a beauty salon because the calculated values of DAEP for workers were very high (5.5–16.6 $mg \cdot (kg\ b.w.)^{-1} \cdot day^{-1}$). To evaluate the air quality and predict exposure effects to VOCs for this scenario, chromatographic analyses of air samples or used products are necessary. For this purpose the methods described in this paper should be helpful.

4 CONCLUSIONS

Analyses of eight nail polish removers and ten nail polishes available on the Polish market were performed.

In the analyzed nail polish removers five volatile organic substances (acetone, ethyl acetate and methyl, ethyl and isopropyl alcohols) were identified. The total VOCs content in analyzed samples varied from 58 to 99%, making nail polish removers a potential source of indoor air pollution.

Emissions of nail polishes usually include aliphatic alcohols and acetates. Some nail polishes might be also sources of aromatic hydrocarbons. In two analyzed samples aromatic hydrocarbons (toluene, ethylbenzene and xylenes) were identified. Nail polishes might influence indoor air quality because they contain large amounts of VOCs (average 70% w/w).

VOCs concentrations during application of the tested products were calculated. In the majority of considered cases the average 1-h concentrations of volatile organic compounds evolved from analyzed products were in the toxic range, so cosmetics used during manicuring were recognized as substantial sources of VOCs emissions. The estimated average inhalation daily intakes (DAEP) of VOCs for adults at home was in the range from 20 to 60 $\mu g \cdot (kg\ b.w.)^{-1} \cdot day^{-1}$. The estimated DAEP of VOCs at a beauty salon varied from 66 to 218 $\mu g \cdot (kg\ b.w.)^{-1} \cdot day^{-1}$ for clients and from 5.5 to 16.6 $mg \cdot (kg\ b.w.)^{-1} \cdot day^{-1}$ for staff.

The described methodology of VOCs determination allows the identification of substances evolved from nail polish removers and nail polishes and to evaluate the risk associated with the use of nail cosmetics.

REFERENCES

Ayoko, G.A. 2004. Volatile organic compounds in indoor environments. In P. Pluschke (ed.), *Indoor air pollution. The Handbook of environmental chemistry* 4F: 1–36. Berlin, Heidelberg, New York: Springer-Verlag.

Guieysse, B., Hort, C., Platel, V., Munoz, R., Ondarts, M. & Revah, S. 2008. Biological treatment of indoor air for VOC removal: Potential and challenges. *Biotechnol. Adv.* 26: 398–410.

Hutter, H.P., Moshammer, H., Wallner, P., Tappler, P. & Kundi, M. 2005. Volatile organic compounds: guidelines from the Austrian Working Group on Indoor Air. *Proceedings: Indoor Air 2005*: 3519–3522.

Huzar, E. & Wodnicka, A. 2009. Analysis of acetone content in nail polish removers. *Polish Journal of Commodity Science (Towaroznawcze Problemy Jakości)* 1(18): 50–58.

Jensen, A.A. & Knudsen H.N. 2006. Total health assessment of chemicals in indoor climate from various consumer product. *Survey of chemical substance in consumer products*. No. 75. Danish Ministry of Environment, Environment Protection Agency.

Jurvelin, J.A. 2003. *Personal exposures to volatile organic compounds and carbonyls: Relationship to microenvironment concentrations and analysis of sources*. Kuopio: National Public Health Institute (Finland).

Kjærgaard, S.K., Mølhave, L., Pedersen, O.F. 1991. Human reactions to a mixture of indoor air volatile organic compounds. *Atmos. Environ.* 25a: 1417–1426.

Kwon, K.-D. & Jo, W.-K. 2007. Indoor emission characteristics of liquid household products using purge-and-trap method. *Environ. Eng. Res.* 12(5): 203–210.

Kwon, K.-D., Jo, W.-K., Lim, H.-J. & Jeong, W.-S. 2007. Characterization of emissions composition for selected household products available in Korea. *J. Hazard. Mater.* 148: 192–198.

Kwon, K.-D., Jo, W.-K., Lim, H.-J. & Jeong, W.-S. 2008. Volatile pollutants emitted from selected liquid household products. *Environ. Sci. Pollut. Res.* 15: 521–526.

Marcinkiewicz-Salmonowiczowa, J. 1995. *Overview of cosmetics chemistry and technology* (in Polish). Gdańsk: Politechnika Gdańska.

Moossavi, M. & Scher, R.K. 2001. Nail care products. *Clin. Dermatol.* 19: 445–448.

Mølhave, L. 1990. Volatile organic compounds, indoor air quality and health. In D.S. Walkinshaw (ed.), *Indoor Air '90. Proceedings of the 5th International Conference on Indoor Air Quality and Climate*, Toronto, Canada, July 29 – August 3, 5: 15–33.

Nazaroff, W.W. & Weschler, Ch.J. 2004. Cleaning products and air fresheners: exposure to primary and secondary air pollutants. *Atmos. Environ.* 38: 2841–2865.

Oppl, R., Neuhaus, T. & Augustin, R., Access: September 2010. *New draft harmonized CEN standard on emissions from construction products – principles common with ISO 16000 and EN 717-1 and differences* (http://product-testing.eurofins.com/media/594542/new%20draft%20harmonized%20cen%20standard%20on%20emissions%20from%20construction%20products.pdf).

Tanaka-Kagawa, T., Uchiyama, S., Matsushima, E., Sasaki, A., Kobayashi, Hiroshi., Kobayashi, Hiromi., Yagi, M., Tsuno, M., Arao, M., Ikemoto, K., Yamasaki, M., Nakashima, A., Shimizu, Y., Otsubo, Y., Ando, M., Jinno, H. & Tokunaga, H. 2005. Survey of volatile organic compounds found in indoor and outdoor air samples from Japan. *Bull. Natl. Inst. Health Sci.* 123: 27–31.

U.S. Environmental Protection Agency. 1997. *Exposure factors handbook. Volume I. General factors.* Washington: National Center for Environmental Assessment, Office of Research and Development.

Wang, S., Ang, H.M. & Tade, M.O. 2007. Volatile organic compounds in indoor environment and photocatalytic oxidation: State of the art. *Environ. Int.* 33: 694–705.

Wolkoff, P. & Nielsen, G.D. 1996. A new approach for indoor climate labeling of building materials – emission testing, modeling, and comfort evaluation. *Atmos. Environ.* 30(15): 2679–2689.

Wolkoff, P. & Nielsen, G.D. 2001. Organic compounds in indoor air – their relevance for perceived indoor air quality? *Atmos. Environ.* 35: 4407–4417.

Zabiegala, B., Partyka, M. & Namiesnik, J. 2003. Indoor air quality – analytics and monitoring (in Polish). In J. Namiesnik, W. Chrzanowski & P. Szpinek (eds), *New horizons and challenges in environmental analysis and monitoring*: 540–562. Gdansk: CEEAM.

Zhu, J., Cao, X.-L. & Beauchamp, R. 2001. Determination of 2-butoxyethanol emissions from selected consumer products and its application in assessment of inhalation exposure associated with cleaning tasks. *Environ. Int.* 26: 589–597.

Management of Indoor Air Quality – Dudzińska (ed)
 ISBN 978-0-415-67266-5

Improvement of thermal comfort and the possibility of energy savings using personalized ventilation

M. Chludzińska
Air Condition and Heating Department, Faculty of Environmental Engineering, Warsaw University of Technology, Warsaw, Poland

A. Bogdan
Department of Ergonomics, Central Institute for Labour Protection – National Research Institute, Warsaw, Poland

ABSTRACT: Personalized Ventilation (PV) is a system that allows thermal conditions in small localized zones close to each occupant, to be individually controlled by them. PV provides energy-saving potential by expanding the room temperature comfort limits using the ability to create a controlled microenvironment. This paper describes the impact of temperature and direction personalized airflow on the human thermal sensation in different ambient temperatures. The results show that extending the upper room temperature limit to 26–28°C in summer and lowering it to 19–20°C in winter yields energy savings using PV. Furthermore, expanding the room temperature limits does not decrease the thermal sensation or comfort of the occupants.

Keywords: ventilation, personalized ventilation, energy savings, thermal comfort, thermal manikin

1 INTRODUCTION

Effective reduction of energy consumption in ventilation and air conditioning systems leads not only to the improvement of the efficiency of facilities, but also to reduction of the required heating/cooling capacity in a building. Personalized ventilation allows us to create optimum thermal parameters locally, in the closest environment to the occupants, while at the same time eliminating the necessity to maintain determined parameters across the whole room. This solution, potentially, allows us to reduce the required heating/cooling energy and additionally increases comfort, adapting local air parameters to the current needs of each occupant and improving the quality of air the occupant breathes.

Analyses have shown that application of a general air conditioning system assisted by a PV system not only improves perceived quality of indoor air and thermal sensation but also makes it possible to reduce the energy used for preparation of air by 15–30% (Sekhar 2005b). Moreover, PV may be a good solution to reduce the energy consumed for building operation. According to "Energy Directive 2002/91/EC" and PN-EN 15251 standard, energy assessment of buildings must be performed in respect to indoor ambient conditions. Application of the PV system combined with low-emission construction materials and an effective air cleaning method, offers a possibility to reduce running costs of building operation, ensuring comfortable conditions of thermal microclimate and quality of air in the breathing zone. Reduction of energy required for HVAC systems combined with PV systems is possible if we apply some additional strategies, i.e. extend indoor air temperature range using local cooling/heating at the work place (Bauman 1993, Sekhar 2003, 2005a), limit the inflow of fresh air by supplying the air directly into the occupant's breathing zone (Bolashikov 2003, Faulkner 2004, Melikov et al. 2002, Sekhar 2003, 2005a) and by supplying air exclusively when the occupant is present at the work place (Bauman 1993, Seem 1992). As shown in the analyses (Chludzińska 2008) and test results (Schiavon 2008a, 2008b), application of PV, by

Figure 1. Testing stand with built-n PV system used in this study.

decreasing the required volume of ventilating air flow and increasing the ambient temperature in the summer season, allows us to reduce the required cooling capacity by up to 50%.

2 MATERIALS AND METHODS

For the purpose of this study, results available in the literature were analyzed. It was assumed that reduced energy consumption for preparation of air conditioning system is possible by extending the range of indoor air temperatures commonly accepted as comfortable (ASHRAE 2004, PN-B-03421:1978, PN-EN ISO 9920:2007). Ensuring an optimum level of thermal sensation in such conditions will be possible thanks to the application of PV.

In order to investigate the possibilities of reduction of the required heating/cooling capacity in the building by means of PV, a series of experiments was performed, which involved a 16-segment thermal manikin and a group of volunteers.

Measurements performed with the thermal manikin and the volunteers were carried out in the same conditions, however at different PV settings and in different ambient temperatures. Surveys carried out among the participants of the experiment allowed us to evaluate the subjective thermal sensation of an occupant under controlled indoor parameters. Thanks to the application of the thermal manikin it was possible to verify the PV impact on the process of heat exchange between an occupant and his/her environment and to classify the predicted thermal sensation in relation to the comfort zones, as defined by Nilsson (Nilsson 2004).

2.1 *Experimental facilities*

In the course of measurements, both the volunteers and the thermal manikin were placed on the testing stand with an in-built PV system (Fig. 1).

The testing stand was composed of a desk with in-built air devices (ADs) directing the prepared air onto the face or ankles of the occupant. ADs were fitted in the central part of the desk, at the level of the head and ankles of the occupant. Rectangular ventilation grilles with movable horizontal louvers were used. The louvers made it possible to change the direction of air horizontally within a range of ±45°. The initial velocity measured at the AD outlet was approx. 0.8 m/s, while the

Table 1. Temperature and direction of personalized air in particular test variants.

Conditions	Direction	Ambient temperature t_a, °C	Temperature of supplied air (PV), t_n, °C	Conditions	Direction	Ambient temperature t_a, °C	Temperature of supplied air (PV), t_n, °C
Winter	Face (F)	18	22	Summer	Face (F)	26	22
			24				24
		20	21				25
			22			28	24
			24				26
	Ankles (A)	18	22				27
			24		Ankles (A)	26	22
		20	21				24
			22				25
			24			28	24
							26
							27

velocity of air measured along the axis of air flow at the edge of the desk (0.5 m away from the outlet) i.e. in an area where the air flow reaches the sitting occupant, was approx. 0.4 m/s. A detailed description of the testing stand is given in (Bogdan 2010).

In order to maintain constant indoor air parameters, stable in time, the tests were performed in a climatic chamber, 2.5 × 3.0 × 3.0 m, in which a typical office environment was simulated. For detailed control of modelled environment conditions, an additional set of microclimate sensors was applied.

Fresh air was supplied to the climatic chamber exclusively by PV. Continuous ventilation air flow, i.e. 20 l/s, was applied.

The volume of the supplied air was measured by means of a sharp edge orifice made as per ISO 5221 standard (PN-ISO 5221:1994).

Twenty subjects (male university students), participated in the experiment. Each of them participated in one experiment daily. Before each measurement the participants were asked to adapt their clothing to the current conditions. They had no possibility to change it during the measurement. Due to the natural adaptation of the human body to the change of seasons during the year, experiments with the participation of volunteers were carried out in two stages, i.e. in winter, conditions were simulated in February and in summer, conditions simulated in May.

The 16-segment thermal manikin worked in "comfort" mode, for which the surface temperature and heat loss from each body segment were measured. In accordance with the simulated conditions, the manikin was dressed in suitable "summer" or "winter" of determined insulation properties. Summer clothes had a total value of insulation parameters of 1.0 clo, while the "winter clothes" had a total value of insulation parameters of 1.1 clo, as per (PN-EN ISO 9920:2007).

2.2 *Experimental conditions*

Measurements were performed in two series, i.e. in winter and summer conditions. In winter conditions the ambient temperature was maintained at 18°C and 20°C while in summer conditions it was kept at 26°C and 28°C. During the measurements stable relative humidity parameters of ambient air, i.e. 50% were maintained.

Twelve measurement series were performed in summer conditions and 10 in winter conditions. Depending on the adopted variant of the test, temperature, airflow direction of personalized air (face or ankles) and ambient temperature were modified as per Table 1.

Moreover, in order to estimate the effects of PV operation, an additional series of measurements was carried out without PV, in ambient temperature of 18, 20°C in winter and of 26, 28°C in summer, in which only the thermal manikin was used.

Table 2. Single experiment design.

Time:	20 min.	20 min.	10 min.
Place:	Outside climatic chamber	Inside climatic chamber	
Activity	Acclimatization of newly arrived volunteers	Acclimatization of volunteers to chamber conditions	1. ASHRAE's seven-point sensation scale for whole body 2. Local thermal discomfort assessment 3. Assessment of preferences

Table 3. Whole body thermal sensation scale according to ASHRAE.

+3	hot
+2	warm
+1	warm, but comfortable
0	neutral (optimal)
−1	cool, but comfortable
−2	cool
−3	cold

2.3 *Measuring procedure*

Participants in the tests assessed thermal conditions inside the climatic chamber on evaluation sheets concerning local and whole body thermal sensation. In accordance with the design of each measurement (Tab. 2), newly arrived volunteers were allowed to acclimatise for 20 min. to achieve an activity corresponding to office work. Next, the volunteer was asked to sit down at the testing stand in the climatic chamber. After 20 min. of adaptation to ambient conditions, the participants filled in a questionnaire concerning subjective thermal sensation.

The questionnaire consisted of a 7 – grade scale of whole body thermal sensation as per ASHRAE ($+3 \div -3$), and evaluation of local thermal discomfort, where the participants were asked to assess their comfort (+) or discomfort (−) on individual parts of the body. In ASHRAE's scale the thermal sensation of the whole body was as shown in Table. 3.

2.4 *Criteria of assessment*

In order to correctly assess PV effects and to compare individual settings of the system, the equivalent temperature index (t_{eq}) was applied. This index is applied in the assessment of thermal sensation in the conditions of dry heat exchange between a person and their environment (Dufton 1929, 1936, Madsen 1984). This parameter, which takes into account not only ambient temperature but also radiant temperature and air velocity allows us to compare the results of each variant of the test carried out in different temperature conditions and with a different direction of airflow.

In the case of measurements performed with the thermal manikin as a human simulator, t_{eq} value is calculated as follows:

$$t_{eq} = t_s - 0{,}155 I_{cl} \cdot q_T \tag{1}$$

where t_s = surface temperature of the manikin, I_{cl} = thermal insulation of clothing, q_T = heating capacity supplied to the manikin working in "comfort" mode.

Equivalent temperature (t_{eq}) was applied in the analysis of the results of similar tests, e.g. (Cermak 2004, 2007).

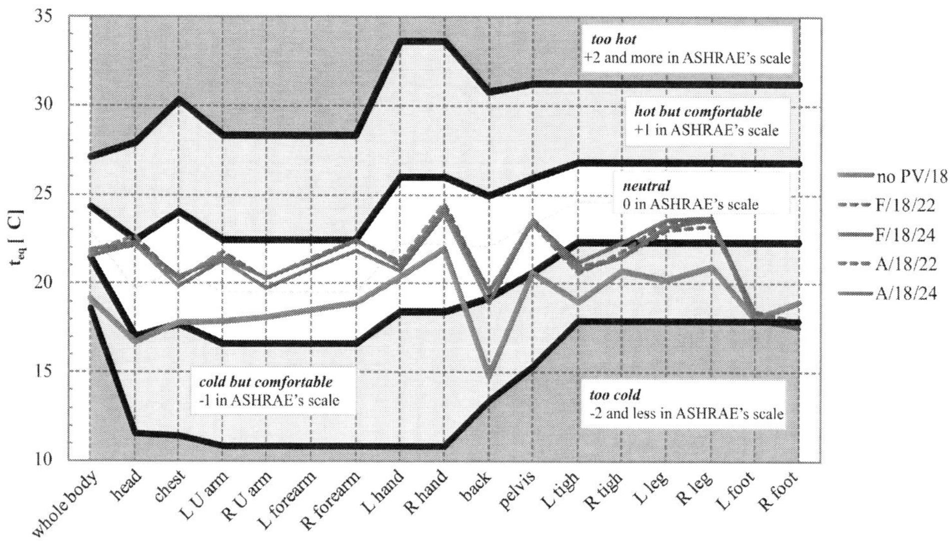

Figure 2. Comfort zones and t_{eq} value for clothed manikin in ambient temperature 18°C (winter conditions 1,6 clo).

3 RESULTS AND DISCUSSION

Figures 2–5 show the results of measurements carried out for the thermal manikin in winter and summer conditions in accord with comfort zones determined by Nilsson (Nilsson 2004). The results shown in figures as F are measurements with personalized air supplies aimed at the face; A are measurements with personalized air supplies pointed at the ankles.

Figure 2, describing winter conditions with variants in ambient temperature of 18°C, shows that application of PV in the temperature of air inflow higher than 4–6K, made it possible to obtain better thermal sensation and maintain the parameters at the comfort level (neutral/optimal) in almost all segments of the manikin. The only exception were the manikin's feet, for which – regardless of PV operation – thermal sensation, as per Nilsson scale, remained at the level between comfortably cool and too cold.

Similar results were obtained in measurement variants carried out at an ambient temperature of 20°C – Figure 3.

Thanks to the PV system it was possible to obtain – on each individual segment – t_{eq} value corresponding to the range of comfortably neutrally, as per Nilsson's comfort zones, with the exception of the feet, for which t_{eq} – both in PV and non-PV variants – corresponded to comfortably cool. However, the PV system at $t_a = 20°C$ did not raise the value of equivalent temperature to the degree obtained in the variants carried out at an ambient temperature of 18°C. The difference in equivalent temperature values for $t_a = 18°C$ reached 5K for the head and the back, while for $t_a = 20°C$ it did not exceed 3K. At the same time, both in $t_a = 18°C$ and in $t_a = 20°C$, the PV system ensured very similar t_{eq} values. This means that it is probably possible to lower the ambient temperature (up to 18°C) slightly while the PV operates, which may decrease the required heating capacity in the winter season.

Analogous observations were made in the summer period. In the measurements carried out at 26°C (Fig. 4) the results obtained remained at the level of neutrally and slightly warm. Application of localized air inflow onto the face lowered the t_{eq} value slightly on upper segments of the manikin, and thus it reached the range *neutral.*

Maintaining the ambient temperature at 28°C (Fig. 5) for the upper segments of the manikin, regardless of the applied combination of PV, made it possible to achieve the range slightly warm and was placed slightly higher than in the measurement variants performed in the temperature of 26°C. No differences, however, were observed for lower segments, which in each case reached the range *neutral.*

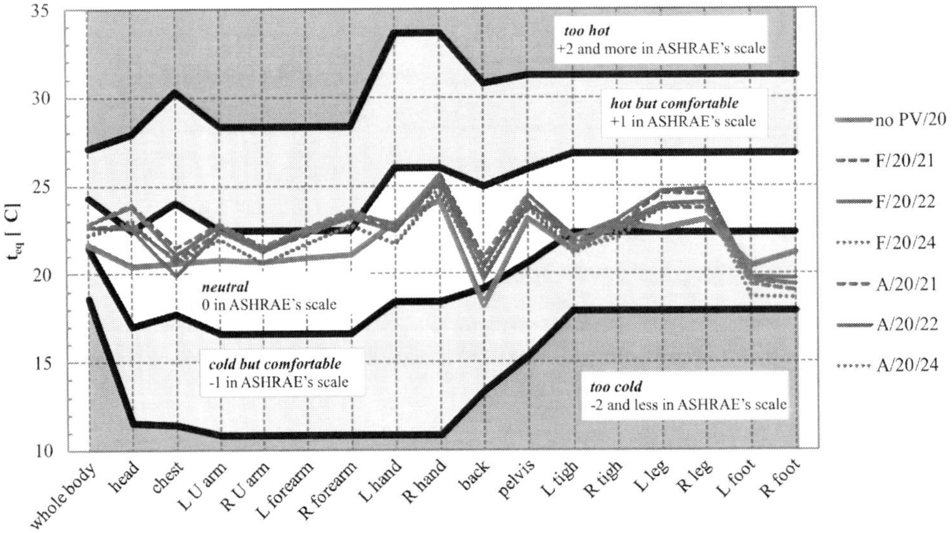

Figure 3. Comfort zones and t_{eq} value for clothed manikin in ambient temperature 20°C (winter conditions 1,6 clo).

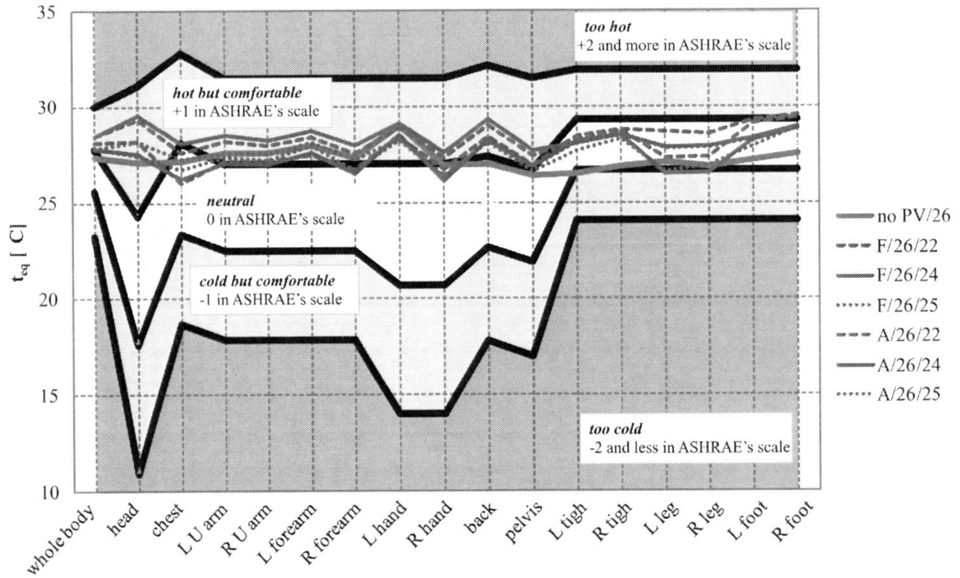

Figure 4. Comfort zones and t_{eq} value for clothed manikin in ambient temperature 26°C (summer conditions 1,1 clo).

Analysis of the results of measurement series with the participation of volunteers has shown that in winter conditions, regardless of ambient temperature in which the test variants were performed and regardless of the temperature and direction of air inflow from the PV system, 80–100% participants declared sensation of thermal comfort, i.e. their responses did not exceed the range of $-1 \div +1$ as per ASHRAE scale. However, 30–75% participants felt comfortable at 0 level, i.e. neutral (Fig. 6). The lowest number of participants declaring a comfortable sensation was recorded for an ambient temperature of 18°C and air inflow temperature of 22°C. However, this number still remained at a high level, i.e. above 80%.

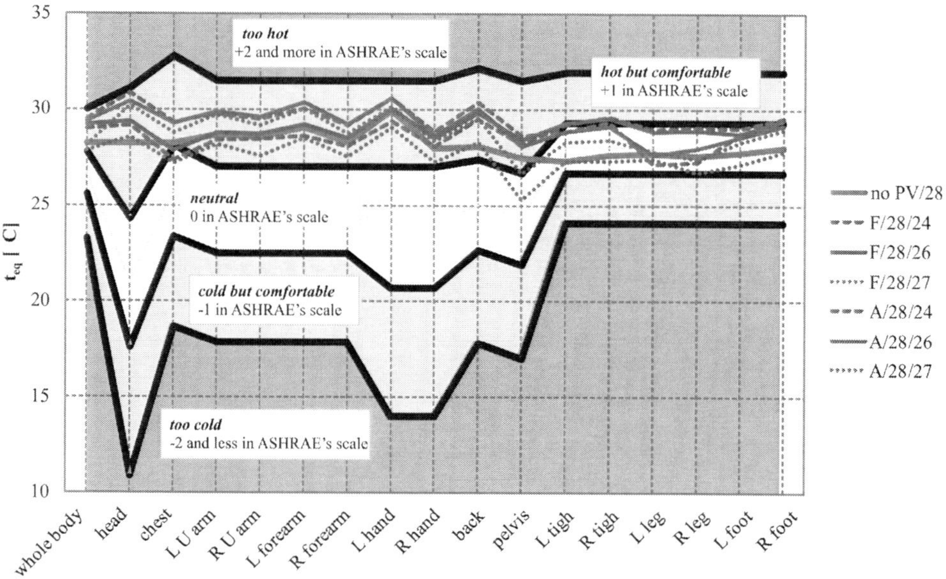

Figure 5. Comfort zones and t_{eq} value for clothed manikin in ambient temperature 28°C (summer conditions 1,1 clo).

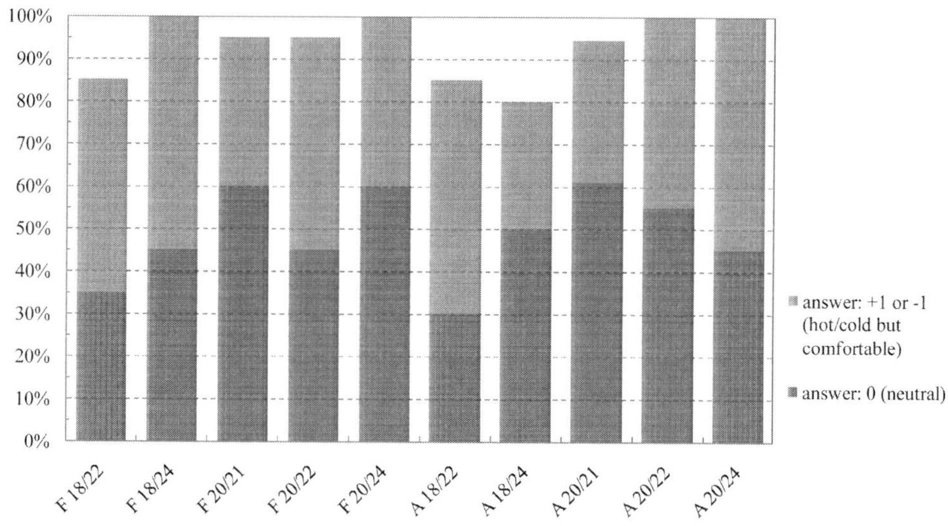

Figure 6. Percentage of subjects reporting thermal comfort of whole body using PV – winter conditions.

In an ambient temperature of 20°C, regardless of the direction of air inflow, the number of participants declaring comfortable thermal sensation was recorded within the range of 95–100%. However, the number of responses at the neutral level (0) increased in comparison with the variants in ambient temperature of 18°C, and reached approx. 55%. During the air inflow onto the face, no correlation in the change of the number of participants responding at 0 level was recorded, however, such a correlation appeared during the air inflow onto the ankles, where, with the increase of temperature of inflow air, the number of participants declaring neutral sensation decreased from 60% to 45%.

Also, in the analysis of thermal sensation on individual parts of the body a large number of participants declared a comfortable thermal sensation on all parts of the body (Fig. 7).

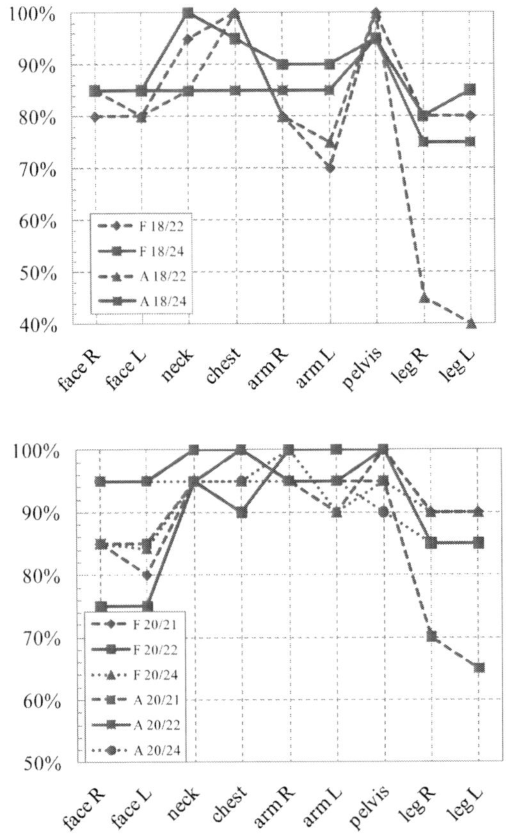

Figure 7. Percentage of subjects reporting thermal comfort on particular body parts in ambient temperature: a) 18°C; b) 20°C using PV – winter conditions.

A slightly lower number of participants declared a comfortable sensation on the face and legs, however, this number still remained at a high level of approx. 75%.

Analysis of the number of people declaring comfortable thermal sensation during the tests performed in the summer season (Fig. 8) has shown a decrease in comfortable thermal sensation with the rise of ambient temperature and air flow, regardless of the direction of airflow.

The highest values were recorded for an ambient temperature of 24°C regardless of the airflow direction and for an ambient temperature of 26°C in the case of airflow onto the face, where the number of participants declaring a comfortable sensation – responses within the range of $-1 \div +1$ – was above 90%. In other cases this number was lower, i.e. approx. 55%. A similar correlation was observed when analysing the number of participants declaring comfort at neutral level in the case of airflow directed onto the face. This number decreased with the increase of ambient temperature and airflow within the range of 60% for F 24/24 variant, and up to the value of 25% for F 28/27 variant. The variants in which airflow was directed onto the ankles, were perceived as less comfortable. The number of participants declaring their sensation at the level of neutral was fixed at approx. 25%. In comparison with the measurement variants performed in winter conditions, the thermal sensation of the whole body in a higher ambient temperature, as compared with the value set in PN-B-03421 standard, i.e. 28°C, was slightly less comfortable in comparison with the measurement variants carried out at lower ambient temperature, i.e. 18°C.

Analysis of the results for individual parts of the body (Fig. 9) has shown that in ambient temperature of 26°C they did not exceed the range $+0.1 \div +1.3$. However, the values above +1, corresponding to the zone comfortably warm, were recorded only on the face and neck, while the values for the other segments did not exceed +0.65.

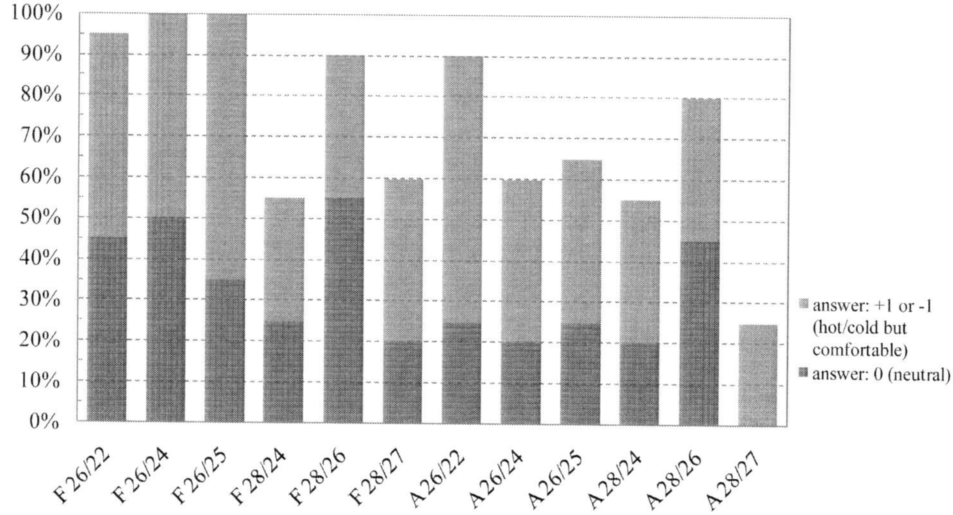

Figure 8. Percentage of subjects reporting thermal comfort of whole body using PV – summer conditions.

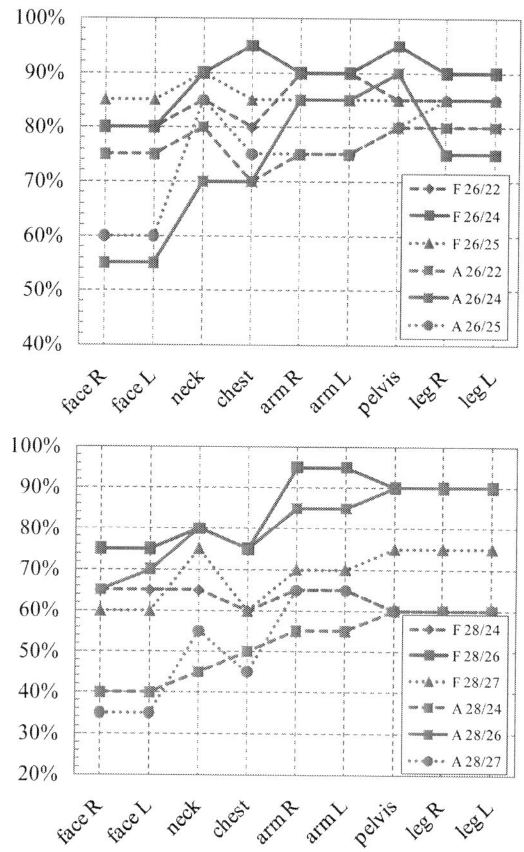

Figure 9. Percentage of subjects reporting thermal comfort on particular body parts in ambient temperature: a) 26°C; b) 28°C using PV – summer conditions.

Additionally, lower values were recorded on the segments exposed to the direct airflow from the PV system. In the case of measurements carried out at 28°C the respondents evaluated their thermal sensation at the level of $+0.6 \div +1.65$. However, again, the maximum values were those of the face and neck during the airflow from the PV system directed onto the ankles. In the case of the other segments, the responses totalled approx. 0.85, and so they did not exceed the range of comfortably warm.

Similarly to the case of ambient temperature of 26°C, lower values were recorded for the parts of the body exposed to the direct airflow from the PV system.

Comparing the results of the measurements using the manikin with the responses given by the volunteers during the tests, especially in ambient temperatures of 26 and 28°C, we can conclude that the value of equivalent temperature at the border of the zones comfortably neutral and comfortably warm is too low.

4 CONCLUSIONS

The tests performed confirmed that it is possible to reduce energy consumption for air conditioned rooms and, at the same time, to improve thermal comfort thanks to the application of the PV system. Personalized ventilation systems, by locally shaping air parameters in the work place, allow us to reduce the energy consumption necessary to maintain suitable temperatures across the whole room. The experimental results have shown that in winter conditions the PV system can be used to warm the human body. To achieve this the supplied air should be directed at the ankles and face, and the temperature should be at least 22°C. To cool the human body in summer conditions, the personalized air should be directed at the face, and the temperature of the supplied air should be lower than the ambient temperature by 2K. In the winter the range of t_{eq} values obtained with the use of the PV system are between "neutral" and "hot, but comfortable", t_{eq} values without the PV air supply stay within the limit of "cold but comfortable". During the summer condition t_{eq} values that result from use of the PV system and air supplied to the face are close to the response of "hot, but comfortable", whereas the values for the air supplied to the ankles and without personalized air supply range between "hot but comfortable" and "too hot".

In the light of the above findings, we conclude that the application of the PV system allows us to maintain ambient temperature at the level of 26–28°C in summer and at the level of 19–20°C in winter, without worsening thermal conditions in the room. This phenomenon potentially allows us to reduce the demand for the cooling and heating capacity. In a moderate climate, much greater benefits can be obtained in the summer season, where application of the PV system allows us to reduce the required difference between indoor and outdoor air by approx. 50–60%. Additionally, thanks to the higher temperature of the supplied air, it also potentially allows us to reduce the sensible cooling capacity, whose value depends on our choice of air conditioning system.

Apart from the benefits related to the reduction of energy consumption, the PV system, combined with fresh air inflow onto the face, allows for highly effective ventilation, improving at the same time the quality of air we breathe.

ACKNOWLEDGEMENTS

The research was financed from funds of the Ministry of Science and Higher Education for the years 2008–2010 as a research project.

REFERENCES

ASHRAE. 2004. Thermal environmental Conditions for Human Occupancy, Atlanta, GA, American Society of Heating, Refrigerating and Air Conditioning Engineers (ASHRAE Standard 55-1992).

Bauman, F.S., Zhang, H., Arens, E.A. & Benton, C.C. 1993. Localized Comfort Control with a Desktop Task/Ambient Conditioning System: Laboratory and Field Measurements. *ASHRAE Transactions* 99 (Pt. 2).

Bogdan, A. & Chludzińska, M. 2010. Assessment of thermal comfort using personalized ventilation. *HVAC&R Research* 16(4): 529–544.

Bolashikov, Z., Nikolev, L., Melikov, A.K., Kaczmarczyk, J. & Fanger P.O. 2003. Personalized ventilation: air terminal devices with high efficiency. *Proceedings of Health and Buildings*, Singapore: 850–855.
Cermak, R. & Melikov, A.K. 2004. Design strategies for personalized ventilation, PhD thesis, International Center for Indoor Environment and Energy, Department of Energy Engineering, Technical University of Denmark.
Cermak, R. & Melikov, A.K. 2007. Protection of Ocupants from Exhaled Infectious Agents and Floor Material Emissions in Rooms with Personalized and Underfloor Ventilation. *HVAC&R Research* 13(1): 23–38.
Chludzińska, M. & Bogdan, A. 2008. Wentylacja indywidualna – kaprys użytkownika czy uzasadniona ekonomicznie oszczędność. *Cieplownictwo, Ogrzewnictwo, Wentylacja* 2(455): 2008.
Dufton, A. F. 1929. The eupatheostat. *Journal of Scientific Instruments* 6: 249–251.
Dufton, A. F. 1936. The equivalent temperature of a warmed room. *JIHVE* (obecnie *Journal of CIBSE*), Wielka Brytania 4: 227–229.
Directive 2002/91/EC of the European Parliament and of the Council of 16 December 2002 on the energy performance of buildings, 16.12.2002.
Faulkner, D., Fisk, W. J., Sullivan, D.P. & Lee S.M. 2004. Ventilation efficiencies and thermal comfort results of a desk- edged-mounted task ventilation system. *Indoor Air* 14(Suppl. 8): 92–97.
Madsen, T., Olesen, B. & Kristensen, N. 1984. Comparison between operative and quivalent temperature under typical indoor conditions. *ASHRAE Transactions* 90(1): 1077–1090.
Melikov, A. K., Cermak, R. & Majer, M. 2002. Personalized ventilation: evaluation of different air terminal devices. *Energy and Buildings* 34: 829–836.
Nilsson, H. 2004. Comfort climate evaluation withthermal manikin methods and computer simulation models, Ph.D Thesis, Department of Civil and Architectural Engineering, Royal Institute of Technology, Sweden.
PN-B-03421:1978 Ventilation and air conditioning. Calculated parameters for indoor air.
PN-EN 15251:2007 Indoor environmental input parameters for design and assessment of energy performance of buildings addressing indoor air quality, thermal environment, lighting and acoustics.
PN-EN ISO 7730:2006 Ergonomics of the thermal environment – Analytical determination and interpretation of thermal comfort using calculation of the PMV and PPD indices and local thermal comfort criteria.
PN-EN ISO 9920:2007 Ergonomics of the thermal environment – Estimation of thermal insulation and water vapour resistance of a clothing ensemble.
PN-ISO 5221:1994 Air distribution and air diffusion – Rules to methods of measuring air flow rate in an air handling duct.
Schiavon, S. & Melikov, A. 2008a. Energy saving and improved comfort by increased air movement. *Energy and Buildings* 40: 1954–1960.
Schiavon, S. & Melikov, A.K. 2008b. Energy saving and improved comfort by increased air movement. *Energy and Buildings* 40: 1954–1960.
Seem, J.E. & Braun, J.E. 1992. The Impact of Personal Environmental Control on Building Energy Use, *ASHRAE Transactions* (Pt. 1): 903–909.
Sekhar, S.C., Gong, N., Cheong, K.W.D., Tham, K.W., Melikov, A.K., Wyon, D.P. & Fanger P.O. 2005a. Findings of Personalized Ventilation Studies in a Hot and Humid Climate. *Heating, Ventilating, Air-Conditioning and Refrigerating Research Journal* 11(4): 603–620.
Sekhar, S.C., Gong, N., Maheswaran, C.R.U., Cheong, K.W.D., Tham, K.W., Melikov, A.K. & Fanger P.O. 2003. Energy efficiency potential of personalized ventilation system in the tropics. *Proceedings of Healthy Buildings*, Singapore 2: 686–689.
Sekhar, S.C., Gong, N., Tham, K.W., Cheong, K.W., Melikov, A.K., Wyon, D.P. & Fanger P.O. 2005b. Findings of Personalized Ventilation Studies in a Hot and Humid Climate. *International Journal of Heating, Ventilation, Air-Conditioning and Refrigerating Research* 11(4): 603–620.

Management of Indoor Air Quality – Dudzińska (ed)
 ISBN 978-0-415-67266-5

Panel heating systems while thermal comfort and energy saving in residential building

M. Rosiński
Faculty of Environmental Engineering, Warsaw University of Technology, Warsaw, Poland

T. Cholewa
Faculty of Environmental Engineering, Lublin University of Technology, Lublin, Poland

Z. Spik
Faculty of Environmental Engineering, Warsaw University of Technology, Warsaw, Poland

A. Siuta-Olcha
Faculty of Environmental Engineering, Lublin University of Technology, Lublin, Poland

ABSTRACT: In general, more efficient heating/cooling technologies are preferred over traditional ones, especially when they are able to maintain a higher level of thermal comfort for occupants in the area. One such option might involve panel heating systems. This paper presents a review of thermal comfort parameters such as: temperature of the heating surface, vertical temperature profile in the room and asymmetry of radiation temperature, for different kinds of panel heating systems. The case of bare feet in contact with the heated floor is introduced. Results indicate that the most suitable finishing material for the radiant floor surface is wood. Attention is also paid to divergences arising from the location of the reference element by calculation of the asymmetry of the radiation temperature. Energy savings and improvement of thermal comfort compared to conventional heating systems are analyzed and discussed.

Keywords: panel heating system, panel cooling system, thermal comfort, energy savings

1 INTRODUCTION

In recent years, a tendency to reduce the projected thermal load in newly designed and modernized residential buildings in Europe has evolved. The trend arose mainly by the necessity to save energy, as laid down in the new Directive 2010/31/UE. This situation has contributed to a growth of interest in low-temperature heating systems, in which the heating medium is supplied at temperatures <55°C. Heating media for this purpose are generally low-temperature heat sources such as a condensing boiler, or heat pumps, which use renewable energy forms.

Among the heat emitters, most widely used are radiant panel heating systems, particularly floor, wall and ceiling heating, which in the previous decade had already been implemented in more than 50% of newly designed buildings in Europe (Kilkis et al. 1994, Laouadi 2004). Nowadays, following from Directives of the European Union, especially those on the limitation of energy consumption and the requirement to increase the proportion of renewable energy use in the residential sector, an intensification of research is clearly evident. This research is directed at optimizing the design stage and controlling radiant panel heating systems in both functions (heating and cooling) by maintenance of thermal comfort conditions (Olesen 2002, Simmonds 1994, Song 2005, Song et al. 2008, Lim et al. 2006). For this reason, the authors decided to carry out a review of the energy saving potential and improvement of thermal comfort conditions arising from the use of radiant panel heating systems in heating and cooling function (Section 2 and Section 3). A comparison with other technical systems commonly used in the heating systems of buildings, is also presented.

2 PANEL HEATING SYSTEM AND THERMAL COMFORT

The heating of closed rooms in cool seasons and their cooling in warm seasons is designed to assure thermal comfort of the room's occupants. This means that a person present in the room will feel neither too cold nor too hot – just comfortable.

For this reason panel heating or cooling systems must assure an acceptable thermal environment both in consideration of thermal comfort parameters connected with the environment (temperature of the heating surface, vertical temperature gradient, asymmetry of radiation, draught) as well as parameters of thermal comfort connected with the human aspect (metabolic transformation level and the clothes). PMV and PPD indicators introduced in ANSI/ASHRAE 55 (ANSI/ASHRAE Standard 55 2004) and ISO 7730 (ISO 7730 2005), used in the evaluation of the thermal comfort level, are usually applied as the tools applicable to the integration of above parameters connected with the environment and personal aspects (Conceição et al. 2010). The requirements of thermal comfort depending on the rooms category (A, B or C) are given in the standard ISO 7730 (ISO 7730 2005).

For rooms of the category A, the standard defines high requirements, for the category B average requirements, while for rooms of the category C–moderate requirements apply. In addition, for every room category the standard fixes the acceptable percentage of persons unhappy with thermal conditions perceived by the total body (PPD) and the acceptable percentage of persons perceiving local discomfort, caused by the four factors shown in Table 1.

As the indicator for thermal comfort used during our research into panel heating systems the applied operative temperature was used. This defines the resultant thermal influence of environments by radiation heating. In a room with air velocity lower than 0.2 m s^{-1}, and difference between mean radiant temperature and air temperature less than 4K, the operative temperature can be calculated as the adjusted air temperature according to ISO Standard 7730 and 7726 and ASHRAE Standard 55. This was used by (Díaz & Cuevas 2010, Fonseca et al. 2010) as a resultant temperature.

When optimizing panel heating systems it should be remembered, that a condition of general thermal comfort is the preservation of the thermal balance of the organism, which is necessary for the maintenance of the constant body temperature. Thus, the body temperature of a healthy person must be approximately constant at about 37°C. Hence, the heat produced by the body as the result of metabolic activity must be returned to the environment. Otherwise this heat would cause an increase in body temperature. In contrast, if the environment constantly absorbs more heat from the organism or over a longer period of time than the person produces, then the mean body temperature will drop. As a result of this, people are very sensitive to changes of temperature, even a slight rise or fall of temperature causing an unpleasant of feeling hot or cold.

Human thermal balance is defined as the situation in which the environment absorbs from the human organism just that amount of heat which the person produces. This can be expressed mathematically in the form of Equation 1.

$$Q_{MET} - Q_D - Q_{evap} - Q_J - Q_U - Q_W = Q_{cond} + Q_K + Q_R \qquad (1)$$

where: Q_{MET} = heat produced in the organism (W); Q_D = heat loss as a result of water vapour diffusion by the skin (W); Q_{evap} = heat loss as a result of vaporizations of the sweat from the

Table 1. Thermal comfort indicators on the basis of the ISO 7730 standard (ISO 7730 2005).

	Thermal sense of the body		Local discomfort			
				PD (%)		
Room category	PPD (%)	PMV	DR (%)	Vertical air temperature difference	Hot, cold floor	Radiation temperature asymmetry
A	<6	−0,2 < PMV < 0,2	<10	<3	<10	<5
B	<10	−0,5 < PMV < 0,5	<20	<5	<10	<5
C	<15	−0,7 < PMV < 0,7	<30	<10	<15	<10

skin surface (W); Q_J = sensible heat loss during respiration (W); Q_U = latent heat loss during respiration (W); Q_W = effective mechanical power (W); Q_{cond} = heat transmitted by conduction (W); Q_K = heat transmitted by convection (W); Q_R = heat transmitted by radiation (W).

Heat exchange by conduction takes place only when the room inhabitant touches the surrounding object e.g. the floor. Theoretical considerations usually treat the person in a standing position, touching the floor only by contact with the feet. Therefore, the heat loss by conduction (Q_{cond}) is a small value, which could be omitted.

Thus, Equation 1 may be written in the form of Equation 2.

$$Q_{MET} - Q_D - Q_{evap} - Q_J - Q_U - Q_W = \alpha_{K-clothes} \cdot A \cdot (t_{clothes} - t_i) + \alpha_{R-clothes} \cdot A \cdot (t_{clothes} - t_{ef}) \tag{2}$$

where: $\alpha_{K-clothes}$ = convective heat transfer coefficient for the surface of the inhabitant's clothes ($W/(m^2 \cdot K)$); A = surface of human body (m^2); $t_{clothes}$ = temperature of clothes surface (°C); t_i = indoor air temperature (°C); $\alpha_{R-clothes}$ = radiant heat transfer coefficient for clothes surface ($W/(m^2 \cdot K)$); t_{ef} = effective temperature of surrounding surfaces, which can be enumerated by use of Equation 4.

In the case of radiation, the decreased value of the surface A, averaging ca. 5–15%, should be applied in the calculation. This is in consideration of the fact that some body parts e.g. hands, trunk or legs shield themselves and "are not seen" by the radiant surface (Cichelka 1965).

2.1 *Temperature profiles by radiant panel heating systems*

Often presented as one of advantages of panel heating, is the resulting uniform air temperature in the room, both in the horizontal and vertical profile. Particularly floor heating produces an air temperature vertical profile closest to an ideal case. This is because of the convective airflow causing compensation of the temperature along the room height and the maintenance of a low value of the temperature gradient.

Research into ceiling heating under static conditions (Rahimi & Sabernaeemi 2010) shows that the vertical air temperature profile in the room, which is 2.5 m in height, can be divided into two zones. The lower zone reaching from the floor to about 1.25 m, in which the air temperature is constant, and the upper zone, in which the difference between extreme points (1.25 m and 2.0 m above the floor) is in the range from 3–5°C depending on the temperature of the heating medium in use.

However, it should be remembered, that a constant air temperature in the heated room is not the same as the uniformity of thermal conditions in the case of radiant heating systems, because they are also defined by the surface effective temperature (see Eq. 4). The distribution of the surface effective temperature is very often not identical to that produced by radiation heating. This inequality can be diminished, by applying a greater heating surface with a lower surface temperature as well as by division of the total heating surface into several smaller surfaces (e.g. the floor heating together with wall heating), which are placed in such a manner that the human body surface is radiated uniformly from different sides.

2.2 *Acceptable surface temperature of radiant panel heating systems*

In case of the radiant heating systems excessive overheating of those body parts nearest the radiant heating sources should be avoided. Unfortunately they are exposed to highest radiation field. For this reason the radiant surface temperature should not exceed a determined allowable limit.

In the case of floor heating, the occupant is constantly in contact with the heating surface, which has a higher temperature than air and other room partitions. Considering that feet are very sensitive to overheating, the floor temperature should not be too high. This places a limit on the thermal performance of radiant floor heating systems. In compliance with the report of CEN Ventilation for Buildings – Design Criteria for Indoor Environment (Raport CEN CR 1752 1998) the maximum acceptable floor temperature amounts to 26°C. In the case of radiant floor heating systems it is allowed to design for a maximum heating surface temperature of 29°C in the people occupancy zone and of 32°C in the boundary zone.

It should be remembered that the values cited above and these listed in Tables 2 and 3, refer to thermal comfort conditions for persons wearing footwear.

Table 2. Acceptable range of floor surface temperatures (Raport CEN CR 1752 1998).

Room category	Floor surface temperature (°C)
A	19–29
B	18–29
C	17–31

where: A – high requirements; B – average requirements; C – moderate requirements.

Table 3. The maximum floor surface temperature (EN1264-2 2008).

Indoor temperature (°C)	Maximum floor surface temperature (°C)	Description
20	29	people occupancy zone
24	33	bathroom and similar
20	35	peripheral area

Table 4. The composition of sensible thermal sensation depending on the coefficient of the heat assimilation.

Heat assimilation coefficient (b)	Thermal sensation
$b \leq 350$	The floor is warm to the feet
$350 < b \leq 700$	The floor is slightly warm or cool to the feet
$b > 1400$	The floor is cold to the feet

In the case of bare feet, not only the floor surface temperature influences thermal comfort, but also the material used for the finishing layer. The physical properties of a finishing material are characterized by the coefficient of heat assimilation (b), which can be calculated by use of Equation 3.

$$b = \sqrt{\lambda \cdot \rho \cdot c_p} \tag{3}$$

where λ = coefficient of heat conduction of floor finishing material (W/(m · K)); ρ = density of floor finishing material (kg/m^3); c_p = specific heat of floor finishing material (J/(kg · K)).

Using the coefficient b, it is possible to evaluate approximately the optimum floor surface temperature from the point of view of thermal comfort. Some model calculations are collected in Table 4.

The positive thermal sensations, caused by the contact of bare feet with the floor surface finished with different materials, were investigated by Strzeszewski (2002). Results of part of this research are presented in Figure 1.

Bearing in mind that the annual surface temperature of a typical radiant floor ranges between 19–32°C, we can conclude that:

- the most suitable finishing material for the radiant floor surface from the point of view of thermal sensations (contact of bare feet with the floor surface), is wood,
- to bare feet in contact with the radiant floor surface, the use of stone or marble as finishing material will result in the sensation of cold if the radiant floor surface temperature is in the range of 28–33°C.

However, in an analysis of the radiant floor calorific effect, the situation reverses, because one of parameters influencing the amount of emitted heat from the heating surface is the temperature of radiant heating surface.

Song (2005) investigated experimentally and Jin et al. (2010), using a computational method, the efficiency of the radiant floor heating in heating/cooling mode by varying the floor finishing

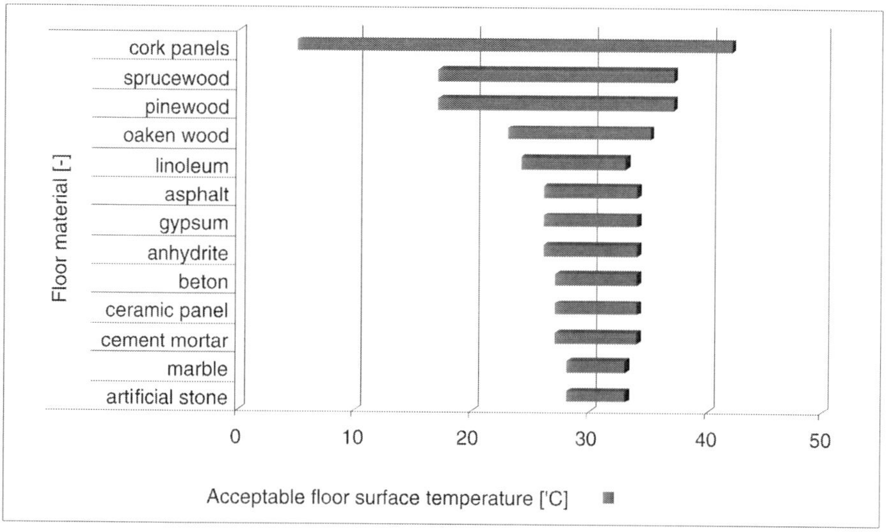

Figure 1. The range of the comfortable temperature of radiant floor surfaces provided by different finishing materials (based on Strzeszewski 2002).

Table 5. The dependence of floor surface temperature (°C) on the temperature of the supply heating medium and finishing material of the radiant surface (modified after Song 2005).

	t_{supply} (°C)					
Material	15	20	25	30	35	40
Mortar	17.0	20.8	24.5	28.1	31.8	35.2
Plywood	19.1	21.3	23.5	26.1	28.1	30.1
PVC	18.2	21.1	24.0	27.0	30.0	32.7
Carpet	18.2	20.5	24.5	26.0	29.5	29.7
Aluminium	–	20.4	24.8	27.7	31.2	34.4

layer and the temperatures of the heating medium supplied to the radiant floor heating system. Song's findings are listed in Table 5.

Analysing data listed in Table 5, it can be seen, that only floors made from wood and its derivatives are characterized with the lower value of the temperature (in heating function) and higher (in cooling function), which decreases the calorific effect of the floor radiant system. Thereby, it should be remembered that, when applying stone or marble as finishing material for the radiant floor heating, (better from the perspective of thermal output than wooden floors), the outer layer temperature of the radiant floor heating should be kept in the range shown in Figure 1. Certainly one should, if possible, avoid carpets as finishing materials of floor heating, because these are also one of main sources of pollution in rooms (Olesen 2002).

In consideration of the fact that the most of heat is exchanged by way of radiation and that persons are close to the heat source (floors), the same panel heating system can be used to cool, which also limits capital investment costs.

In the situation, in which the radiant floor system is supplied with a cooling medium and will contribute to room cooling during Summer, the limiting value of the bottom floor surface temperature should, for rooms of category A and B amount to 19°C. However, for a room of category C it should be 17°C. This fact also influences the ability to obtain the cooling capacity from 1 m^2 of the radiant floor surface in the cooling function.

Wall heating systems can provide higher heating surface temperatures than the ceiling or floor surfaces. This is why the temperature of the wall heating surface with pipes situated inside building

Table 6. Acceptable values of asymmetry of radiation temperature (Glück 1994, Dudkiewicz & Jezowiecki 2007).

Type of panel heating/cooling	Asymmetry of radiation temperature according to Fanger (K)	Asymmetry of radiation temperature according to Glück (K)
Ceiling heating	4	8.1
Wall cooling	10	8.4
Ceiling cooling	14	14.3
Wall heating	23	11.1

partitions could be even in the range 50–55°C. This allows the thermal output of 1 m^2 of the heating wall surface to be enlarged. However one should be aware that the upper edge of the heating panel should not be higher than 1.5 m over the floor surface.

2.3 *Asymmetry of radiation temperature*

The concept of the asymmetric thermal field is variously defined in the literature, because the relevant research was carried out using different measurement methods, different assumptions and measuring conditions. Early research on the topic were carried out by Chrenko in 1953, who arrived at the acceptable temperature of a heated ceiling system as a function of the dimensions of the heating surface and the height of its installation. Fanger, Banhidi, Olesen & Langkilde (Fanger 1974), defined the concept of "asymmetry of radiation temperature" as a result of work of the influence of a heated ceiling on the sensations of persons in sitting position in the climate chamber. The asymmetry of radiation temperature was defined as the difference of the temperature of the two radiation spaces surrounding the surface of a horizontal element situated 0.6 m above the floor (the body centre of the seated persons), parallel to the surface with greatest temperature differences. This definition found wide recognition in the field, but still raises many doubts. Among other things, the location height of the reference element – 0.6 m above the floor, has been questioned (Glück 1991, Glück 1993, Glück 1994). This is connected with the fact that, in rooms in which the occupants are without head cover and other parts are usually protected from the radiation by suitable clothing. That is why, the head is considered as most sensitive, while, on the basis of asymmetry of radiation, the human body element takes on this role. In view of this, Glück (1991, 1994) accepted the height of the head of a sitting male as 1.3 m above the floor, and the standing man at 1.7 m. Comparing acceptable values of the asymmetry of radiation temperature indicated in Table 6, certain divergences between Fanger's and Glück's work are evident.

In the case of ceiling heating room occupants feel radiation on the crown of the head, while below it the remaining body is turned to the cold surrounding surfaces. Thus, from this point of view, the acceptable asymmetry of radiation temperature is at its minimum in this instance. A much more tolerable situation for occupants is given when the surrounding cold surfaces are at headheight, such as occurs when the ceiling is used in the cooling function. In this case the crown of the head and the face return heat in the direction of the cooling surfaces, thus allowing a greater temperature difference in the surrounding surfaces.

2.4 *Comparison of radiant heating with other systems*

Resulting from, among other things, the temperature of the surrounding surfaces, the so-called effective temperature t_{ef} (Eq. 4), being different in the case of radiant heating systems than the air temperature of the room, radiant heating differs considerably from convection heating.

Besides its thermal influence, the radiant heating surface has a directional character, i.e. the heating surface radiates only one side of the human body, thus reducing (with the square of the distance) the radiation intensity.

$$t_{ef} = \sqrt[4]{\varphi_{1,R} \cdot T_1^4 + \varphi_{2,R} \cdot T_2^4 + \ldots + \varphi_{n,R} \cdot T_n^4} - 273.15 \quad (4)$$

where $\phi_{1,R}$, $\phi_{n,R}$ = configuration coefficient between radiant surface F_R and surrounding surfaces F_1, F_n which are in the "view zone" of the surface F_R; T_1, T_n = absolute temperature of the surface F_1, F_n (K).

In the study (Myhren & Holmberg 2008) comparisons were made between low-temperature, medium-temperature and traditional high-temperature systems in regard to assurances of thermal comfort conditions. The analysed systems were:

- radiant floor heating system supplied with low-temperature heating medium (t_{supply}/t_{return}: 35/25°C);
- radiant wall heating system supplied with low-temperature heating medium (t_{supply}/t_{return}: 35/25°C);
- a system with convective radiators supplied with medium-temperature heating medium (t_{supply}/t_{return}: 55/40°C);
- a system with convective radiators supplied with high-temperature heating medium (t_{supply}/t_{return}: 90/70°C).

Results show that, for the radiant floor and wall heating system, the value of the PPD indicator was close to 5% for the most part of occupancy zones. However, in case of convective radiators, the PPD values were higher, a little above 5% in the occupancy zone. In addition, it was observed that convective radiators create acceptable thermal comfort conditions in a part of the occupancy zone located near the wall with the window under which the radiator is situated. One reason is the fact that the convective radiator, which is situated under the window, causes heating up of incoming cold air already close to the window, which has a ventilation supply opening. However, in the case of floor heating, this incoming cold air must enter into direct contact with the radiant heating panel before it can be heated to an acceptable temperature.

During the comparison of heating systems the average air speed which, to prevent the sensation of discomfort among the most sensitive persons, should be <0.15 m/s in the occupancy zone. In the studies (Holmberg et al. 2004, Myhren & Holmberg 2006, Myhren & Holmberg 2008) it could be shown that low-temperature systems are characterized by smaller values of air speed, because the temperature of the heating surfaces is close to that of room air, thus diminishing convective air movements and dust flotation in comparison to high-temperature systems (Olesen 1997, Olesen 2002).

Myhren & Holmberg (2008) demonstrated that low-temperature systems, particularly using radiant panel heating systems, contribute to improvement of the thermal comfort conditions inside rooms, largely due to the decrease in air speed and the temperature difference inside the room, in comparison to conventional high-temperature systems. The improvement of thermal comfort parameters and of the indoor air quality in rooms with low-temperature heating systems was confirmed, also in further studies (Olesen et al. 1980, Hutter 1991, Eijdems & Boerstra 2000, Nilsson 2003).

However, in the study (Imanari et al. 1999) the authors compared a ceiling system with a heating and cooling function with a traditionally installed air-conditioning system. In the case of the ceiling system, the ventilating air volume was limited to the minimum according to sanitary requirements. It was shown that the ceiling system in the heating and cooling function assure smaller differences of the vertical temperature profile as well as being able to create a more comfortable indoor environment (Tab. 7).

On the basis of the results shown in Table 7, it can be confirmed that the ceiling heating is less beneficial with respect to thermal comfort than ceiling cooling. This is presumably because the heads of occupants in the heating function are nearer the ceiling heating surface but the feet are distant. This the reverse of the accepted optimum thermal comfort condition of humans.

3 PANEL HEATING SYSTEMS AND ENERGY SAVING

The use of radiant panel heating systems generates energy savings in comparison to conventional heating systems, as was noticed in early theoretical or experimental studies (Feustel & Stetiu 1995, Yost et al. 1995, Stetiu 1999). Heat transfer by radiation, which is the main mechanism of heat exchange in panel heating systems, to people, surrounding objects and partitions is more economic

Table 7. Results of the estimation of thermal comfort conditions for different systems (Imanari et al. 1999).

Kind of system	Function of system	The estimation of thermal comfort conditions in the room (%)			
		Comfortable	Slightly comfortable	Neutral	Slightly uncomfortable
ceiling	cooling	44.5	33.3	22.2	–
air-conditioning	cooling	45.5	9.0	45.5	–
ceiling	heating	13.6	25.8	53.0	7.6
air-conditioning	heating	4.4	30.4	47.8	17.4

than heating up the total air volume by way of convection, which fills the heated room. This fact allows householders to lower the projected indoor air temperature in winter about 1–2°C, which decreases the energy consumption by circa 10%. Similarly, in summer one can increase the indoor temperature and consequently diminish the demand on the cooling power by maintaining acceptable thermal comfort requirements, in comparison with traditional systems.

In the study (Jin et al. 2010) the influence of reducing the velocity of cooling medium flow in the pipes of the floor system in its cooling function, was investigated. The authors pointed out that the difference between the temperature of floor surface during laminar and turbulent flow in the pipes was only about 0.3°C. For this reason, it is recommended to lower the velocity of medium flow in the pipes of the heating panel. This will diminish the frictional drag and decrease the required elevation height of the circulation pump, which, in turn, will generate energy savings. Similarly in the study (Díaz & Cuevas 2010), it was found that the value of the mass flow rate in the pipes of the ceiling heating, has little influence on heating/cooling efficiency. Hence, it is also recommended to choose the water flow rate carefully, with the aim of reducing the electrical energy consumption of pumps.

Energy savings due to use of a water floor heating system was also observed in the study (Hasan et al. 2009), in which four different heating systems were simulated.

Most disadvantageous with respect to the primary energy consumption was system 1 equipped with convective radiators in rooms supplied with heating medium (70/40°C) and with the electric heating in the bathroom. The remaining three systems were:

- system 2: floor heating in rooms and bathroom with the supply heating medium at 40°C and with the control system based on the indoor temperature (21°C),
- system 3: floor heating in rooms and bathroom with the supply heating medium at 40°C and with the control system based on the floor surface temperature (24°C),
- system 4: heating with convective radiators in rooms with the supply/return heating medium at (45/35°C) and floor heating in the bathroom with the supply heating medium at 45°C.

These three systems consumed about 50% less primary energy than system 1. Besides, it was seen that the control of the floor heating system using the indoor air temperature (system 2) reduced the energy consumption in comparison to the system 3, which is controlled on the basis of a given temperature of the floor heating surface.

In the radiant ceiling heating system, a considerable reduction in the air volume, transported within the building occurred in comparison to normal air-conditioning systems. Often only 20% of the air volume, characteristic for the conventional air-conditioning system, is transported (Conroy & Mumma 2001). Such a reduction can lead to a diminution of fan size and power, electrical energy consumption and the cross-sections of air ducts (Wilkins & Kosonen 1992). Similarly Feustel & Stetiu (1995) demonstrated that use of a radiant panel cooling system could lower the electrical energy consumption by as much as 40% in comparison with HVAC systems in industrial buildings existing at that time.

One other method of saving energy in radiant panel heating is the proper control of their operation (Chen 2002). In the study (Cho & Zaheer-uddin 2003) the authors state that the use of the control method of floor heating using the predictable hourly outdoor temperature, can give energy savings in the range of 10–12% in comparison with standard control methods.

4 SUMMARY

Among the major environmental goals of the EU are the limitation of energy consumption and the increased use of "renewables" in the residential sector. Hence, it appears convincing that radiant panel heating systems, resulting from their high energy savings in comparison to other heating systems, could become one of the primary measures contributing to the improvement of energy efficiency. Furthermore, panel heating systems are in a position to assure a higher level of thermal comfort characterized by thermal comfort parameters (temperature of the heating surface, vertical temperature profile in the room, asymmetry of radiation temperature) in comparison to other heating/cooling systems. Nevertheless, energy savings will not be achievable without a suitable control strategy, a computational model, as well as without the exact designing of the radiant panel heating installation.

ACKNOWLEDGEMENTS

This study was supported by project No. N N523 7452 40, financed by the Polish Ministry of Science and Higher Education.

REFERENCES

ANSI/ASHRAE Standard 55. 2004. Thermal environmental conditions for human occupancy. American Society of Heating, Refrigerating and Air-Conditioning Engineers.

Chen, T.Y. 2002. Application of adaptive predictive control to a floor heating system with a large thermal lag. *Energy and Buildings* 34: 45–51.

Cho, S.H. & Zaheer-uddin, M. 2003. Predictive control of intermittently operated radiant floor heating systems. *Energy Conversion and Management* 44: 1333–1342.

Cichelka, J. 1965. *Heating by radiation*. Warsaw: Arkady.

Conceição, E.Z.E., Manuela, M. & Lúcio, J.R. 2010. Evaluation of thermal comfort conditions in a localized radiant system placed in front and behind two students seated nearby warmed curtains. *Building and Environment* 45: 2100–2110.

Conroy, C. & Mumma, S. 2001. Ceiling radiant cooling panels as a viable distributed parallel sensible cooling technology integrated with dedicated outdoor air systems. *ASHRAE Transactions* 107(1): 578–585.

Díaz, N.F. & Cuevas, C. 2010. Testing and thermal modeling of radiant panels systems as commissioning tool. *Energy Conversion and Management* 51: 2663–2677.

Dudkiewicz, E. & Jeżowiecki, J. 2007. *Research on heat field in halls heated with gas infrared radiators*. Lublin: Monograph of the Environmental Engineering Committee PAN, vol. 45.

Eijdems, H.H.E.W. & Boerstra, A.C. 2000. Low Temperature Heating Systems: Impact on IAQ, Thermal Comfort and Energy Consumption. Annex 37, Newsletter 1.

EN 1264-2. 2008. Water based surface embedded heating and cooling systems – Part 2: Floor heating: Prove methods for the determination of the thermal output using calculation and test methods.

Fanger, P.O. 1974. *Thermal comfort*. Warsaw: Arkady.

Feustel, H.E. & Stetiu, C. 1995. Hydronic radiant cooling—preliminary assessment. *Energy and Building* 22: 193–205.

Fonseca, N., Bertagnolio, S. & Cuevas, C. 2010. Radiant ceiling systems coupled to its environment part 2: Dynamic modeling and validation. *Applied Thermal Engineering* 30: 2196–2203.

Glück, B. 1991. Bewertungsmaßstab zur optimalen Anordnung von Heiz- und Kühlflächen im Raum. *Gesundheits-Ingenieur* (2): 65–71.

Glück, B. 1993. Empfindungstemperaturen und ihre Messbarkeit. *Gesundheits-Ingenieur* (3): 134–142.

Glück, B. 1994. Zulässige Strahlungstemperatur-Asymmetrie. *Gesundheits-Ingenieur* (6): 285–293.

Hasan, A., Kurnitski, J. & Jokiranta, K. 2009. A combined low temperature water heating system consisting of radiators and floor heating. *Energy and Buildings* 41: 470–479.

Holmberg, S., Molin, F. & Myhren, J.A. 2004. Space heating with low temperature difference between heating unit and ambient air. In *Proceedings of the 9th International Conference on Air Distributions in Room*, Portugal, September 2004: 162–163.

Hutter, E. 1991. Comparison of different heat emitters in respect of thermal comfort and energy consumption. In *Proceedings of the International Centre for Heat and Mass Transfer, Heat and Mass Transfer in Building Materials and Structures*: 753–769.

Imanari, T., Omori, T. & Bogaki, K. 1999. Thermal comfort and energy consumption of the radiant ceiling panel system. Comparison with the conventional all-air system. *Energy and Buildings* 30: 167–175.
ISO 7730. 2005. Ergonomics of the thermal environments – analytical determination and Interpretation of thermal comfort using calculation of the PMV and PPD indices and local thermal comfort criteria. International Standard. Switzerland.
ISO 7726. 1998. Ergonomics of the Thermal Environment—Instruments for Measuring Physical Quantities.
Jin, X., Zhang, X. & Luo, Y. 2010. A calculation method for the floor surface temperature in radiant floor system. *Energy and Buildings* 42: 1753–1758.
Jin, X., Zhang, X., Luo, Y. & Cao, R. 2010. Numerical simulation of radiant floor cooling system: The effects of thermal resistance of pipe and water velocity on the performance. *Building and Environment* 45: 2545–2552.
Kilkis, B.I., Sager, S.S. & Uludag, M. 1994. A simplified model for radiant heating and cooling panels. *Simulation Practice and Theory* (2): 61–76.
Laouadi, A. 2004. Development of a radiant heating and cooling model for building energy simulation software. *Building and Environment* 39: 421–431.
Lim, J.H., Jo, J.H., Kim, Y.Y., Yeo, M.S. & Kim, K.W. 2006. Application of the control methods for radiant floor cooling system in residential buildings. *Building and Environment* 41: 60–73.
Myhren, J.A. & Holmberg, S. 2006. Comfort temperatures and operative temperatures in an office with different heating methods. In *Proceedings of the Healthy Buildings, vol. 2: Indoor Climate*, Portugal, June 2006: 47–52.
Myhren, J.A. & Holmberg, S. 2008. Flow patterns and thermal comfort in a room with panel, floor and wall heating. *Energy and Buildings* 40: 524–536.
Nilsson, P.E. 2003. *Achieving the Desired Indoor Climate.* IMI Indoor Climate and Studentlitteratur, ISBN 91-44-03235-8.
Olesen, B.W. 1997. Possibilities and limitations of radiant floor cooling. *ASHRAE Transactions* 103: 42–48.
Olesen, B.W. 2002. Radiant floor heating in theory and practice. *ASHRAE Journal* 44(7): 19–26.
Olesen, B.W., Mortensen, E. & Thorshaug,e J. 1980. Thermal comfort in a room heated by different methods. *ASHRAE Transactions, 86: Technical Paper* no. 2256.
Rahimi, M. & Sabernaeemi, A. 2010. Experimental study of radiation and free convection in an enclosure with a radiant ceiling heating system. *Energy and Buildings* 42: 2077–2082.
Raport CEN CR 1752. 1998. Ventilation for Buildings – Design Criteria for Indoor Environment, European Committee for Standardization.
Simmonds, P. 1994. Control strategies for combined heating and cooling radiant systems. *ASHRAE Transaction* 100: 1031–1039.
Song, D., Kim, T., Song, S., Hwang, S. & Leigh, S.B. 2008. Performance evaluation of a radiant floor cooling system integrated with dehumidified ventilation. *Applied Thermal Engineering* 28: 1299–1311.
Song, G.S. 2005. Buttock responses to contact with finishing materials over the ONDOL floor heating system in Korea. *Energy and Buildings* 37: 65–75.
Stetiu, C. 1999. Energy and peak power potential of radiant cooling systems in US commercial buildings. *Energy and Buildings* 30: 127–138.
Strzeszewski, M. 2002. Computational model of microline-heatings. PhD thesis, Warsaw University of Technology, Warsaw.
Wilkins, C.K. & Kosonen, R. 1992. Cool ceiling systems: a European air conditioning alternative. *ASHRAE Journal* (8): 41.
Yost, P.A., Barbour, C.E. & Watson, R. 1995. An evaluation of thermal comfort and energy consumption for a surface mounted ceiling radiant panel heating system. *ASHRAE Transactions* 101(1): 1221–1235.

Management of Indoor Air Quality – Dudzińska (ed)
 ISBN 978-0-415-67266-5

The choice of an optimal heating system for a low-energy single-family house under Polish climatic conditions – a case study

A. Siuta-Olcha & T. Cholewa
Faculty of Environmental Engineering, Lublin University of Technology, Lublin, Poland

ABSTRACT: In the article the problem of equipping low-energy buildings with highly efficient and low emissive heating installations with the aim of minimizing heating costs is discussed. The subject of the study was a single-family house with defined construction for which the annual primary power demand, the annual final power demand and the heat demand for heating purpose, were determined. The heated floor area of this building is about 198.1 m^2. A special computer program packet dedicated to passive building design (PHPP) was used for computational analysis.

Keywords: low-energy building, energy sources, energy consumption, energy costs

1 INTRODUCTION

Interest in problems of energy-efficient construction in Poland and its practical application arose mainly from a change of energy policy (Directive 31/2010/UE; Dz.U. No 75, item 690, 2002, Dz.U. No 201, item 1238 and item 1240, 2008) and the necessity to act regarding saving energy and improving the environment. The essential feature of energy-efficient buildings is to subject all the detailed solutions to construction and installation requirements of low energy consumption as well as environmental protection requirements. The structure of energy consumption by end use in Polish households for 2002 was: 71.2% for space heating, 15.1% hot water preparation, 6.6% cooking, 4.5% electrical equipment, 2.3% lighting. In European residential buildings about 57% of the total final energy consumption is used for space heating, 25% for domestic water heating and 11% for electricity (Chwieduk 2003, Balaras et al. 2007). The largest share in energy consumption in buildings is space heating. Annual energy demand for heating purposes depends on the standard of the building (Wojdyga 2009). The maximal value of the index for seasonal heat demand for new building standards is 100 kW · h/(m^2 · year), for low-energy buildings 30 kW · h/(m^2 · year), for passive houses 15 kW · h/(m^2 · year) (Balaras et al. 2007, Audenaert et al. 2008, Wojdyga 2009, Mahdavi & Doppelbauer 2010). For all high performance houses a maximum of 60 kW · h/(m^2 · year) of non-renewable primary energy for space heating, hot water heating and electricity for mechanical systems is foreseen (Smeds & Wall 2007). According to (Chan et al. 2010) energy consumption in the building sector constitutes 35.3% of the final energy demand.

The energy demand for space heating in energy-efficient buildings of the latest generation should be only 30% of the total energy load of the building. In the case of the existing buildings, thermal needs for heating purposes are at about 100 W/m^2 whereas for energy-efficient buildings these are within 25–35 W/m^2 (Feist et al. 2006). Excessive energy consumption for heating houses in Poland, is related to: (i) losses of energy through external walls, roof, floor and windows, (ii) poor thermal efficiency of heat sources and elements of heating systems, (iii) lack of measuring devices for consumed heat and the lack of settlement of heating costs among the various recipients, (iv) lack of serious consideration as to the shape of a building and its location in terms of heat losses.

For passive houses the energy demand has been reduced to 10 W/m^2, which is why it is possible to provide heat for space heating only through the existing system of mechanical supply-exhaust ventilation. A passive house is a building with an extremely low energy demand for heating, where thermal comfort is assured by internal heat sources such as the inhabitants, heat loss of electrical appliances, lighting and solar heat gains, heat gains from the ventilation and, in individual cases,

Table 1. U_{max}-values of the building components and the recommended thickness of insulation for three building types.

Building component	Standard building	Low-energy building	Passive house
Exterior walls	$d_{ins} = 12$ cm $U = 0.30$ W/(m$^2 \cdot$ K)	$d_{ins} = 18$ cm $U = 0.18$ W/(m$^2 \cdot$ K)	$d_{ins} =$ min 30 cm $U = 0.15$ W/(m$^2 \cdot$ K)
Roof	$d_{ins} = 15$ cm $U = 0.25$ W/(m$^2 \cdot$ K)	$d_{ins} =$ min 20 cm $U = 0.20$ W/(m$^2 \cdot$ K)	$d_{ins} =$ min 40 cm $U = 0.10$ W/(m$^2 \cdot$ K)
Windows	$U = 1.80/1.70$ W/(m$^2 \cdot$ K)	$U = 1.70$ W/(m$^2 \cdot$ K)	$U = 0.80$ W/(m$^2 \cdot$ K)

by reheating the ventilation air (Badescu & Sicre 2003). The total final energy consumed by a passive building is less than a quarter of the energy consumed by an average new house built according to the requirements of thermal protection of the building. The required standard for a low-energy or passive house can be achieved through the implementation of certain technical solutions. Particular attention should be paid to: (i) adequate thermal insulation of building elements (through conventional thermal modernization demand for heat can be reduced by about 35–40%) and use of special energy-efficient windows, (ii) reduction of thermal bridges, (iii) air tightness of the building envelope, (iv) ensuring effective ventilation with heat recovery from exhaust ventilation air.

The maximum values of overall heat-transfer coefficients through the envelope of residential buildings in Poland are provided in the regulation (Dz.U. No 75, item 690, 2002). The U_{max}-values of the building components and the recommended thickness of insulation for buildings of various standard are shown in Table 1.

The adequate spacing of glass surfaces, choosing the right type and size of windows can significantly reduce the energy demand for space heating. Proper location and design of windows allows for maximum benefits of solar heat in winter and reduction of the intense heating of the interiors in summer. The total solar energy transmission factor of glazing in passive buildings should be taken as 50–60%.

Reducing heat losses by eliminating thermal bridges is practically a difficult task to perform. The linear thermal bridge heat loss coefficient, relative to the external dimensions, should not exceed 0.01 W/(m $\cdot$ K).

Reducing the energy consumption of the building allows the use of high efficiency low-temperature heating systems with convection heaters (for example steel plate radiators) or/and panel radiators (floor heating systems, wall heating systems). The ideal solution, in this case, may be a heat pump or compact heating and ventilation device with a heat pump unit of brine/water type with an integrated domestic hot water tank. Very often, heating systems are designed with a gas condensing boiler or a low-temperature gas boiler (or an oil-fired boiler) with a closed combustion chamber. Fireplaces with a closed hearth reduce the demand for primary energy. One can also consider introducing electric heating of the building.

To determine the energy demand for the preparation of domestic hot water in passive buildings, a level of domestic hot water consumption should be adopted as 25 dm^3/(day $\cdot$ person). In this case, the most commonly used heat sources are two-function gas condensing boilers, heat pump, boilers automatically powered by renewable fuel, for example, wood pellets, electric flow water heaters. All of these sources may cooperate with the installation of liquid solar energy collectors.

In low-energy buildings, natural ventilation is usually replaced by a mechanical supply-exhaust ventilation installation with heat recovery and a ground heat exchanger for preheating the outside air.

2 MATERIALS AND METHODS

2.1 *Building description*

The object of the analysis is a single-family detached house without a basement. The building is occupied by two adults and three children. This building is 16 m in length and 16 m in width. Its height is about 3 m (Fig. 1).

The external walls are constructed with lime-sand blocks SILKA E, 0.24 m wide and with mineral wool as insulating material, 0.3 m wide. The ceilings containing a well-ventilated air layer are

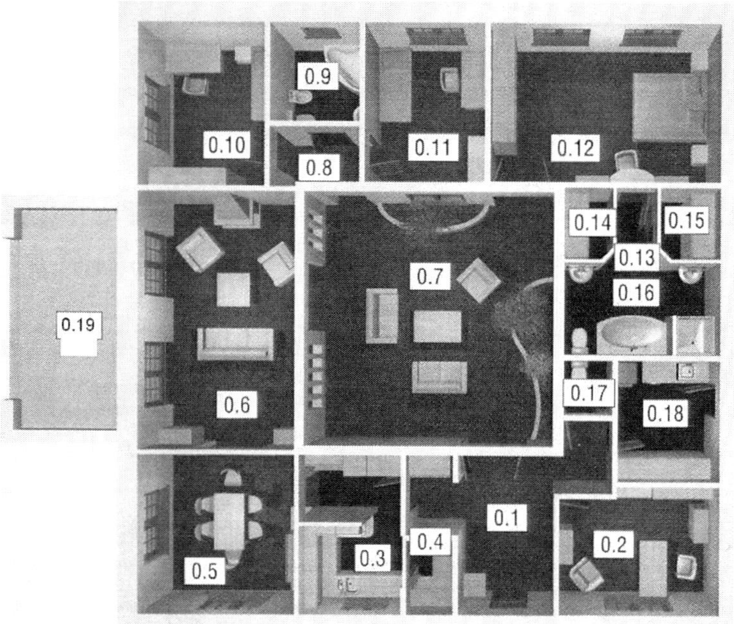

Figure 1. Structure of the analyzed low-energy house: 0.1 corridor and hall (12.49 m^2), 0.2 study (12.75 m^2), 0.3 kitchen (10.29 m^2), 0.4 pantry (2.23 m^2), 0.5 dining room (5.92 m^2), 0.6 drawing room (25.86 m^2), 0.7 atrium (40.45 m^2), 0.8 corridor (5.68 m^2), 0.9 bathroom (5.68 m^2), 0.10 room for children (13.19 m^2), 0.11 room for children (12.21 m^2), 0.12 room for parents (23.58 m^2), 0.13 corridor (1.63 m^2), 0.14 dressing room (2.17 m^2), 0.15 dressing room (2.76 m^2), 0.16 bathroom (9.79 m^2), 0.17 retiring room (1.7 m^2), 0.18 store-room (8.08 m^2), 0.19 terrace.

Table 2. U-values of the analyzed building components.

Building components	U-value, $W/(m^2 \cdot K)$
Exterior wall	0.12
Ceiling over the drawing room	0.16
Ceiling over rooms	0.11
Floor on the ground (without cellar beneath)	0.15
Windows	0.92
Exterior Door	2.10

insulated with 0.34/0.24 meter-thick mineral wool. The floor at ground level is thermally insulated with 15 centimeter-thick extruded polystyrene – there was no underlying cellar under this level. The building is covered with a hip roof. The inclination angle of this roof is given as 23 degrees. The house has a living area of 198.1 m^2 and an inner volume of heated parts of the building of 536 m^3. The area of all windows in this building is 38.85 m^2. The windows have Aluplast profiles with triple glazing. The total solar energy transmittance of the glazing is 0.52. The U-values of the building components are presented in Table 2.

The air tightness of the building was assumed to be equal to 0.5 air changes per hour at a pressurisation of 50 Pa. The building has a mechanical supply-exhaust ventilation system with high efficiency heat recovery from the air being removed without a ground-air heat exchanger. The effective efficiency of the heat exchanger is 85%. Energy consumption by this recuperator is 0.36 $W \cdot h/m^3$. The required exhaust air flow is 180 m^3/h.

The building is equipped with energy-efficient household appliances (dishwasher, washing machine, refrigerator-freezer) and energy-efficient lighting. The building is equipped with a solar

Table 3. Design exterior temperature θ_e and average annual exterior temperature $\theta_{m,e}$ (PN-EN 12831:2006).

Climatic zone	θ_e (°C)	$\theta_{m,e}$ (°C)
I	−16	7.7
II	−18	7.9
III	−20	7.6
IV	−22	6.9
V	−24	5.5

water heating system. The required area of solar vacuum tubes is 4 m^2. The average annual share of solar energy in meeting the thermal needs for hot water generation is 58%. The amount of heat supplied by the collectors was set at 1,858 kW · h/year. The installation of photovoltaic cells is not planned.

2.2 *Weather data*

Poland is located in the northern and eastern hemisphere between 49°00′ and 54°50′ north latitude and between 14°07′ and 24°08′ east longitude. The country is divided into five climatic zones (PS-82/B-02403, PS EN 12831:2006). The values of the external air temperature and average yearly temperature for each zone are shown in Table 3.

The Polish climatic database introduced into the Passive House Planning Package (PHPP) was used in the computer simulations. The temperature of the external air and monthly sums of solar radiation incident on horizontal and vertical surfaces of different orientation was taken into account for each month, for a given location.

2.3 *The Passive House Planning Package (PHPP)*

The PHPP program was developed at the Passive House Institute in Darmstadt in Germany. It was designed specifically for low-energy buildings. The PHPP is used to assess the energy performance of buildings and to determine the total energy costs assuming various heating-system solutions. Application of the PHPP allows, among other things, the calculation of the annual space heating demand, the nominal heat load, the useful heat requirement of domestic hot water, the electricity demand of the building, the final and primary energy demands.

The building energy demand for space heating is calculated as given by Equation 1:

$$Q_{H,nd} = Q_{H,ht} - \eta_{H,gn} \cdot Q_{H,gn} \,, \quad \text{kW·h/year} \tag{1}$$

where:
$Q_{H,ht}$ – the total heat transfer for the building, kW · h/year:

$$Q_{H,ht} = Q_T + Q_V \,, \quad \text{kW·h/year} \tag{2}$$

Q_T – the total heat transfer by transmission, kW · h/year,
Q_V – the total heat transfer by ventilation, kW · h/year:

$$Q_T = \left(\sum_i A_i \cdot U_i + \sum_k l_k \cdot \Psi_k\right) \cdot f_T \cdot G_T \cdot 10^{-3} \,, \qquad \text{kW·h/year} \tag{3}$$

where:
A_i – the area of element i of the building envelope, m^2,
U_i – the thermal transmittance of element i of the building envelope, W/(m^2 · K),
l_k – the length of linear thermal bridge k, m,
Ψ_k – the linear thermal transmittance of linear thermal bridge k, W/(m · K),
f_T – the temperature factor,

G_T – the number of heating degree-hours, K · h/year; $G_T = 89{,}600$ K · h/year, and

$$Q_V = n_w \cdot V_w \cdot c \cdot G_T \ , \quad \text{kW·h/year} \tag{4}$$

n_w – the energy exchange coefficient of air, h^{-1}:

$$n_w = n \cdot (1 - \Phi_R) + n_{\text{inf}} \ , \quad h^{-1} \tag{5}$$

where:
n – average air change rate, $n = 0.3\ h^{-1}$,
Φ_R – efficiency of heat recovery, $\Phi_R = 0.85$,
n_{inf} – infiltration air change rate, $n_{\text{inf}} = 0.037\ h^{-1}$,
V_w – the building ventilation volume, m^3,
c – the heat capacity of air per volume, $c = 0.33$ W · h/(m^3 · K).
$Q_{H,gn}$ – the total heat gains for the building, kW · h/year:

$$Q_{H,gn} = Q_S + Q_i \ , \quad \text{kW·h/year} \tag{6}$$

where:
Q_S – the sum of solar heat gains over the given period, kW · h/year:

$$Q_S = \sum\nolimits_k r_k \cdot g_k \cdot A_{w,k} \cdot S_k \ , \quad \text{kW·h/year} \tag{7}$$

where:
r_k – the reduction factor which takes into account the percentage of the surface of the window frame, $r_k = 0.44$,
g_k – the total solar energy transmittance of the transparent part of glazed envelope elements of surface k, $g_k = 0.52$,
$A_{w,k}$ – the window area of surface k, with a given orientation, m^2,
S_k – the global solar radiation of surface k, with a given orientation, kW · h/(m^2 · year) and
Q_i – the sum of internal heat gains over the given period, kW · h/year:

$$Q_i = 0.024 \cdot z_o \cdot q_i^j \cdot A_f \ , \quad \text{kW·h/year} \tag{8}$$

where:
z_o – length of the heating season, d/year,
q_i^j – the heat flow rate from internal heat sources, W/m^2,
A_f – the heating surface of the building, m^2.

The dimensionless gain utilization factor for heating $\eta_{H,gn}$ can be expressed by the following formula:

$$\eta_{H,gn} = \frac{1 - \left(\dfrac{Q_{H,gn}}{Q_{H,ht}} \right)^5}{1 - \left(\dfrac{Q_{H,gn}}{Q_{H,ht}} \right)^6} \tag{9}$$

The specific space heat demand is determined as follows:

$$E_{H,nd} = \frac{Q_{H,nd}}{A_f} \ , \quad \text{kW·h/(m}^2\text{·year)} \tag{10}$$

The total building need for domestic hot water is calculated in accordance with Equations 11–12: kW · h/year:

$$Q_{T,HW} = Q_{HW,nd} + \Delta Q_1 + \Delta Q_2 \,, \qquad \text{kW·h/year} \tag{11}$$

$$Q_{HW,nd} = q^j_{HW} \cdot U \cdot c_w \cdot \rho_w \cdot (t_{HW} - t_{CW}) \frac{365}{3600} \,, \qquad \text{kW·h/year} \tag{12}$$

where:
q^j_{HW} – daily hot water consumption of per inhabitant; $q^j_{HW} = 25$ dm^3/d per inhabitant,
U – the number of inhabitants,
c_w – specific heat of water; $c_w = 4.19$ kJ/(kg · K),
ρ – density of water; $\rho = 1$ kg/dm^3,
t_{HW} – domestic hot water temperature, °C,
t_{CW} – cold water temperature, °C,
ΔQ_1 – the average annual loss of heat transport system of hot water, kW · h/year,
ΔQ_2 – the average annual loss of heat hot water tank, kW · h/year.

The annual electricity demand for electrical equipment (or auxiliary electricity) is given by Equation 13, kW · h/year:

$$E_{el} = \sum V_{norm} \cdot f_u \cdot h \cdot G \cdot f_{el} \tag{13}$$

V_{norm} – normative power consumption by electrical devices, kW · h/use or kW · h/d,
f_u – utilization factor (correction factor for the normative power consumption),
h – frequency of use during the year in relation to the norm electricity demand,
G – reference quantity for h,
f_{el} – electric fraction – percentage of electricity demand.

The total final energy demand is based on Equation 14:

$$Q_{F,H+W} = e_1 \cdot Q_{H,nd} + e_2 \cdot Q_{HW,nd} \,, \qquad \text{kW·h/year} \tag{14}$$

where:
e_1 – performance ratio of heat generator space heat run,
e_2 – performance ratio of heat generator domestic hot water run.

The specific final energy demand is determined as follows:

$$EK_{H+W} = \frac{Q_{F,H+W}}{A_f} \,, \qquad \text{kW·h/(m}^2\text{·year)} \tag{15}$$

The total primary energy demand is calculated by Equation 16:

$$Q_P = \sum p \cdot Q_F \,, \qquad \text{kW·h/year} \tag{16}$$

where:
p – the coefficient of non-renewable energy source of primary energy.

The specific primary energy demand is given by Equation 17:

$$EP = \frac{Q_P}{A_f} \,, \qquad \text{kW·h/(m}^2\text{·year)} \tag{17}$$

The equivalent emission factor CO_2 can be expressed by the following formula:

$$u_{CO2} = \frac{x_{CO2} \cdot E_{(H+HWD)}}{A_f} \,, \qquad \text{kg/(m}^2\text{·year)} \tag{18}$$

where:
x_{CO2} – the equivalent emission factor CO_2, kg/(kW · h),
$E_{(H+HWD)}$ – the annual emissions equivalent, kg/year.

Table 4. Tariff groups for consumers of natural gas (values without VAT) (www.pgnig.pl).

Tariff group	Annual fuel consumption, m^3/year	Natural gas price, PLN/m^3	Network charge (fixed rate), PLN/month	Network charge (variable rate), PLN/m^3	Subscription charge, PLN/month
W-1	≤800	0.998	3.85	0.6029	4.30
W-2	800÷1200	0.984	11.05	0.4795	7.05
W-3	1200÷8000	0.970	43.35	0.3542	8.20

Table 5. Tariff group G11 for Polish consumers of electricity (values without VAT) (www.pgesa.pl).

Price of electricity, PLN/(kW · h)	0.2483
Network charge (fixed rate)	
– one phase system,	1.90
– three phase system, PLN/m-c	2.94
Network charge (variable rate), PLN/(kW · h)	0.2201
Qualitative rate, PLN/(kW · h)	0.0077
Transient charge for annual electricity consumption:	
– <500 kW · h,	0.38
– 500÷1200 kW · h,	1.59
– >1200 kW · h, PLN/m-c	5.03
Subscription charge, PLN per two months	3.30

Table 6. Design space heating demand of the analyzed single-family house as a function of meteorological conditions.

Climatic zone	I	II	III	IV	V
Annual energy demand for space heating, kW · h/(m^2 · year)	24	26	29.1	31	33
Heating load, W/m^2	16.4	21.8	22.4	22.7	23.2

2.4 *Fuel costs*

To determine the energy costs for the house, appropriate prices for fuel and charge rates, the subscription and network charge, based on existing tariffs, were assumed. In Table 4 depicts the unit fees for the supplied natural gas (group E), depending on annual consumption of this fuel (excluding VAT). The calorific value of natural gas is taken as 36 MJ/m^3. A statement of prices and rates for electricity users according to the group G11 tariff is given in Table 5.

In the case of the use of the low-temperature oil boiler, the energy cost was defined assuming the price of light heating oil at 2.79 PLN/dm^3 and a calorific value of 41.86 MJ/kg. When considering a heat source in the form of biomass-combusting boiler, the price of wood of an average calorific value of 12 MJ/kg, was assumed to be 180 PLN/m^3.

3 RESULTS AND DISCUSSION

The energy demand was calculated using the PHPP program. Energy balances of the analyzed single-family building were prepared for all the highlighted climatic zones. For the calculation of internal heat gains a standard rate of 2.1 W/m^2 was assumed. In Table 6 the values of annual energy demand for space heating and heating load for this building, are presented in dependence of its location.

Detailed assessment of the energy performance of the building together with the determination of the costs of used energy was performed using the PHPP for the third climatic zone (III). The

Table 7. Monthly heat losses and heat gains balance for the analyzed building.

Month	Heat losses of the building, kW · h/month	Solar heat gains, kW · h/month	Internal heat gains, kW · h/month	Specific space heating demand, kW · h/(m^2 · month)
January	1954	185	310	7.4
February	1701	331	280	5.5
March	1594	553	310	3.7
April	1120	757	300	0.5
May	709	1020	310	0
June	429	925	300	0
July	321	1001	310	0
August	355	904	310	0
September	638	644	300	0
October	1031	488	310	1.2
November	1376	241	300	4.2
December	1752	130	310	6.6
Year	12981 kW · h/year	7179 kW · h/year	3644 kW · h/year	29.1 kW · h/(m^2 · year)

heating load of the house was calculated to be 4,434 W (22.4 W/m^2) and the yearly energy demand for heating purposes as 5,765 kW · h (29.1 kW · h/m^2) which is below the limit of 30 kW · h/m^2 per year allowed by low-energy house requirements. Use of a solar hot water system allows for the reduction of heat demand for hot water preparation from 3,186 kW · h/year to 1,328 kW · h/year.

The following options for heating system and preparation of domestic hot water and cooking were assumed for the analysis:

- Case 1: The heat source is a two-function low-temperature gas boiler with an efficiency of 90%; cooking – electric stove,
- Case 2: The heat source is a two-function low-temperature gas boiler with an efficiency of 90%, cooking – 4-burner gas stove,
- Case 3: The heat source is a two-function gas condensing boiler with an efficiency of 99%, cooking – electric stove,
- Case 4: The heat source is a two-function gas condensing boiler with an efficiency of 99%, cooking – 4-burner gas cooker,
- Case 5: The heat source is a two-function low-temperature oil boiler with an efficiency of 90%, cooking – electric stove,
- Case 6: The heat source is a compact heating and ventilation unit with heat pump (Coefficient of performance in heating mode COP = 4.5) and with an integrated hot water tank, cooking – electric stove,
- Case 7: The heat source is a two-function wood-burning boiler, with an efficiency of 80%, cooking – electric stove.

The analysis does not include heating up space with a fireplace heater.

The values of gains and losses of heat in the building for each month are shown in Table 7. The highest total heat losses (by transmission and by ventilation) occur in January.

The amount of final energy required to meet the energy demand of the building depends on the efficiency of the system. The annual demand for electricity and final energy for space heating purposes and hot water preparation for the considered cases is presented in Table 8.

Figure 2 presents the values of the primary energy demand *EP*(2) including space heating, domestic hot water heating, electricity for fans, the heat exchanger and pumps and the values of total primary energy demand *EP*(1) (also with household electricity) for the all highlighted cases.

To determine the energy costs for the house, the appropriate prices were assumed for fuels and charge rates, the subscription and network charge, based on existing tariffs. Table 9 presents the annual costs of the delivered fuel, electricity costs and the total cost of the energy consumed in the analyzed building for the highlighted options. Costs of hot water preparation and space heating using natural gas as primary energy source were calculated at 0.05 €/(kW · h), using oil at

Table 8. The annual demand for electricity and final energy for space heating and hot water preparation for the considered cases.

Option	Electricity demand, kW · h/year	Total final energy demand (for space heating and domestic hot water) kW · h/year	(EK_{H+W}, kW · h/(m^2 · year))
C1, C5	3094	8403	(42.4)
C2	2594	8403	(42.4)
C3	3094	7756	(39.1)
C4	2594	7756	(39.1)
C6	4598	1574	(7.9)
C7	3001	11860	(59.9)

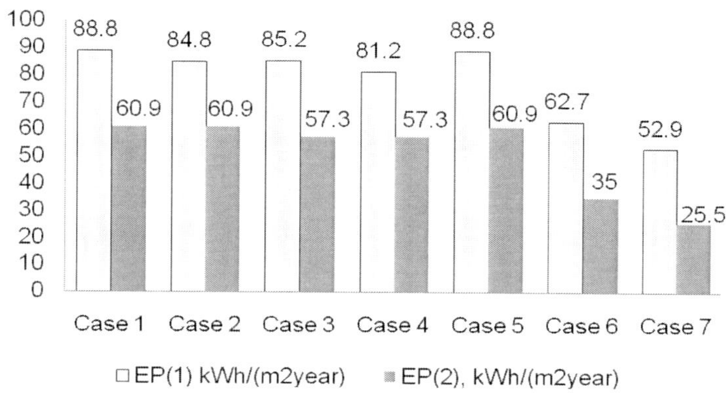

Figure 2. Annual *EP*-values for the analyzed building for different types of energy sources.

Table 9. Yearly costs of the delivered fuel, electricity costs and the total cost of the energy consumed in the analyzed building.

	Fuel cost		Electricity cost		Total energy cost for the building	
Option	PLN/year	(€/year)	PLN/year	(€/year)	PLN/year	(€/year)
C1	1771.27	(440.75)	1917.50	(477.13)	3688.77	(917.89)
C2	1860.90	(463.05)	1627.10	(404.87)	3488.00	(867.92)
C3	1655.18	(411.86)	1917.50	(477.13)	3572.68	(888.99)
C4	1744.81	(434.16)	1627.10	(404.87)	3371.91	(839.03)
C5	2416.93	(601.40)	1917.50	(477.13)	4334.43	(1078.54)
C6	0.00	(0.00)	2791.06	(694.50)	2791.06	(694.50)
C7	798.88	198.79	1863.53	(463.70)	2662.41	(662.49)

0.07 €/(kW · h) and fuel wood of 0.017 €/(kW · h). For comparison, in Belgium the energy cost for heating was calculated by using a mean gas price of 0.04 €/(kW · h) (Audenaert et al. 2008).

The total emissions CO_2 equivalent – values for the analyzed system solutions in the building are shown in Figure 3.

4 CONCLUSIONS

Designing ecological and energy-efficient buildings is a complex process which requires interdisciplinary links between the architectural and urban and the technological and ecological requirements.

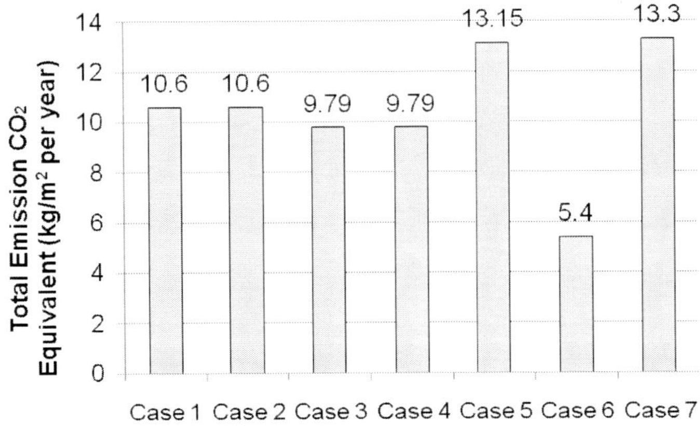

Figure 3. CO_2 equivalent emissions of the delivered energy.

Both reduction of primary energy consumption and the use of renewable energy sources have a significant impact on reducing the operating costs of the building. On the basis of computer simulations we could show that the best heating-system solution for the analyzed building is either the system with a compact brine/water heat pump with solar connection or a biomass boiler. The choice of heating system and source of energy is very important for environmental effects.

REFERENCES

Audenaert, A., De Cleyn, S.H. & Vankerckhove, B. 2008. Economic analysis of passive houses and low-energy houses compared with standard houses. *Energy Policy* 36(1): 47–55.

Badescu, V. & Sicre, B. 2003. Renewable energy for passive house heating. Part I. Building description. *Energy and Buildings* 35(11): 1077–1084.

Balaras, C.A., Gaglia, A.G., Georgopoulou, E., Mirasgedis, S., Sarafidis, Y. & Lalas, D.P. 2007. European residential buildings and empirical assessment of the Hellenic building stock, energy consumption, emissions and potential energy savings. *Building and Environment* 42(3): 1298–1314.

Chan, H.Y., Riffat, S.B. & Zhu, J. 2010. Review of passive solar heating and cooling technologies. *Renewable and Sustainable Energy Reviews* 14(2): 781–789.

Chwieduk, D. 2003. Towards sustainable-energy buildings. *Applied Energy*. 76(1–3): 211–217.

Directive 2010/31/EU of the European Parliament and of the Council of 19 May 2010 on the energy performance of buildings.

Feist, W., Munzenberg, U., Thumulla, J. & Darup, B.S. 2006. *Passive houses*. Schlagowski G. Polish Passive House Institute, Gdansk (in Polish).

Mahdavi, A. & Doppelbauer, E. 2010. A performance comparison of passive and low-energy buildings. *Energy and Buildings* 42(8): 1314–1319.

PHPP 2006: Passive House Planning Package 2006 Manual + CD-Rom (in Polish).

Polish regulation: Decree of the Ministry of Infrastructure, Dz.U. No 75, item 690, 12.4.2002 with changes: No 33, item 270, 2004, No 109, item 1156, 2008, No 201, item 1238.

Polish regulation: Decree of the Ministry of Infrastructure, Dz.U. No 201, item 1238 and item 1240, 6.11.2008.

PS-82/B-02403 Heating. Exteriors calculated temperatures.

PS EN 12831:2006 Heating systems in buildings – Method for calculation of the design heat load.

Smeds, J. & Wall, M. 2007. Enhanced energy conservation in houses through high performance design. *Energy and Buildings* 39(3): 273–278.

Wichowski, R. 2007. Energy consumption in residential buildings for selected European countries. *District Heating, Heating, Ventilation*, 11(452): 70–71 (in Polish).

Wojdyga, K. 2009. An investigation into the heat consumption in a low-energy building. *Renewable Energy* 34(12): 2935–2939.

www.pgesa.pl (30.08.2010).

www.pgnig.pl (30.08.2010).

Management of Indoor Air Quality – Dudzińska (ed)
ISBN 978-0-415-67266-5

Thermal conditions and aerosol particle number concentration in a lecture room

B. Połednik & M.R. Dudzińska
Faculty of Environmental Engineering, Lublin University of Technology, Lublin, Poland

ABSTRACT: The results presented in this paper deal with the effects of thermal conditions on the aerosol particle number concentration in a lecture room. The influence of naturally changed indoor air temperature and humidity on particle concentration was examined. Measurements of particle number concentrations and thermal conditions were performed in a vacant lecture room and then again when the students were present. The obtained results suggest that indoor air thermal parameters, except occupancy and outdoor concentration of particles, could influence the number concentration of aerosol particles in the lecture room. The presence of students caused an enormous increase in coarse particles (>5 μm) and a significant decrease of small particles (<5 μm). These changes could have implications for students exposed to aerosol particles. There are possible associated health consequences, especially when these particles are potentially hazardous.

Keywords: aerosol particles, indoor air temperature, relative humidity, specific enthalpy, classroom environment

1 INTRODUCTION

The influence of temperature and humidity fluctuations on the content of aerosol particles in indoor air has still not been satisfactorily explained (Gorbunov et al. 1999, Kramer et al. 1999, Park et al. 2002). The mechanism of these aerosol changes is even more unclear when the indoor air temperature and humidity fluctuations are caused by the presence of people. The infiltration of outdoor aerosols to the room is different when people are present. Breathing and human activity are also substantial factors which cause changes in the concentration of aerosol particles (Nazaroff 2004, Połednik & Dudzińska 2010). People generate significant amounts of indoor aerosols. These aerosols potentially have an important influence on short-term expositions (Jankowska et al. 2004, Afshari et al. 2005).

Contemporary research has mainly been concerned with fine and ultrafine indoor aerosol particles. However, the available information on coarse particles, both in terms of their indoor behavior and the health effects they cause, is still relatively limited (Morawska et al. 2003). The toxic, carcinogenic, inflammatory, allergenic and other adverse properties of these particles is still not well understood. This lack of understanding arises from the fact that this particulate matter is composed of different types of particles (i.e. soil minerals, biological particles, fibers of various origin, semi-volatile and non-volatile organics), and synergic reactions may also occur (Braniš et al. 2005). It is not entirely clear how the number concentration of coarse particles is associated with indoor activities and microclimatic changes. The concentration of these particles is reduced relatively fast by such processes as filtration, gravitational sedimentation and deposition onto indoor surfaces (Hussein et al. 2005). Some of these particles could also make their way into the human respiratory track. According to the health risk investigations (EUR 19529/EN 2000), particles smaller than 2.5 μm may penetrate deep into the lung and about half of them are not exhaled. In turn, bigger particles can be deposited at differentiated degree in the trachea, bronchi and upper respiratory track.

Generally it is assumed that aerosol particles, especially coarse particles (on account of their bigger mass, volume and surface area) could affect the abundance of trace elements and chemical compounds in the indoor environment. When health effects are considered, the properties and behavior of indoor coarse particles and dust are very important (Toivola et al. 2004, Reff et al. 2005). Overall, scientific evidence suggests that this particulate matter, in combination with other factors can cause respiratory and cardiovascular diseases when deposited in the human respiratory tract (e.g. Steinemann 2004, Pöschl 2002 and references therein).

This paper presents the measurement results of the number concentration of aerosol particles (>0.3 μm) in a vacant and occupied lecture room. The results take into account the naturally changing indoor air thermal parameters.

2 SITE AND METHODS

The aerosol particle number concentration, indoor air temperature and relative humidity measurements were performed in one of the lecture rooms of the Lublin University of Technology (LUT) in Poland. The lecture room was located on the second floor of the five-storied LUT building. The 86 m^3 cuboid room had two double paned windows and was gravitationally ventilated. Inside there was a lector's desk, tables and seats for 24 students, one blackboard for chalk writing, and one whiteboard. The room was additionally equipped with a slide and a data projector. The number of students (aged between 18 and 22) participating in the experiment varied from a few to enough students for full room capacity. The lecture room occupancy usually changed every second hour according to the traditional academic system of 2-h lectures. Separate lectures were divided by 15 minute breaks during which the students moved in and out, and the room was ventilated by opening windows. The lecture room was irregularly occupied due to the schedule of the LUT courses.

The participating students were fully informed about the experiment. Size and concentration of the aerosol particles that where sampled in the lecture room were determined by means of a laser particle counter ROYCO 243A. The samples were collected at 30 second intervals with a 15 minute delay time. The counter categorized the collected particles into four size ranges: 0.3–0.5, 0.5–5, 5–10 and >10 μm.

The particle concentration measurements relied on the continuous registration of particle number concentration changes. They were performed in an alternately vacant and occupied lecture room, at naturally changed indoor thermal conditions. Parallel measurements of indoor and outdoor particle concentrations were unworkable. All measurements in the investigated lecture room were fulfilled during the 12 weeks of the spring semester (Połednik & Dudzińska 2008).

3 RESULTS AND DISCUSSION

Changes in the number concentration of selected sizes of aerosol particles, air temperature and relative humidity in the lecture room without and with the presence of students are shown in Figure 1. The diagrams illustrate the summed up changes caused by both fluctuations of indoor and outdoor aerosols, as well as the presence of students in the examined lecture room (Lai 2002, Holmberg & Chen 2003). Because information of outdoor particle concentrations is not available, possible influence can only be estimated. The changes of particle number concentrations in the lecture room without students and in the absence of other indoor particle sources (Fig. 1a), seem to be mainly caused by fluctuations in outdoor aerosols (Asmi et al. 1999, Franck et al. 2003). However, taking into account that the examined lecture room had a fairly air-tight door and windows which restricted the inflow of outdoor aerosols, other factors such as indoor air thermal parameters should also be considered (Połednik et al. 2007). The alternating indoor air temperature and humidity inside the lecture room could affect the number concentrations of the smallest particle measured (0.3–0.5 μm). Indeed, from the diagrams in Figure 1, it is visible that the highest amplitude changes concern particles of these sizes.

The diagrams in Figure 1b were plotted only for the intervals when students were present in the room. Therefore, it can be assumed that the apparent changes of particle number concentrations are mainly caused by the presence and mobile activity of students. These factors could be considered as a substantial source of bulky aerosol particles in a lecture room (Połednik & Dudzińska 2008).

Table 1 presents basic statistical information about the number concentration of the measured particle size fractions.

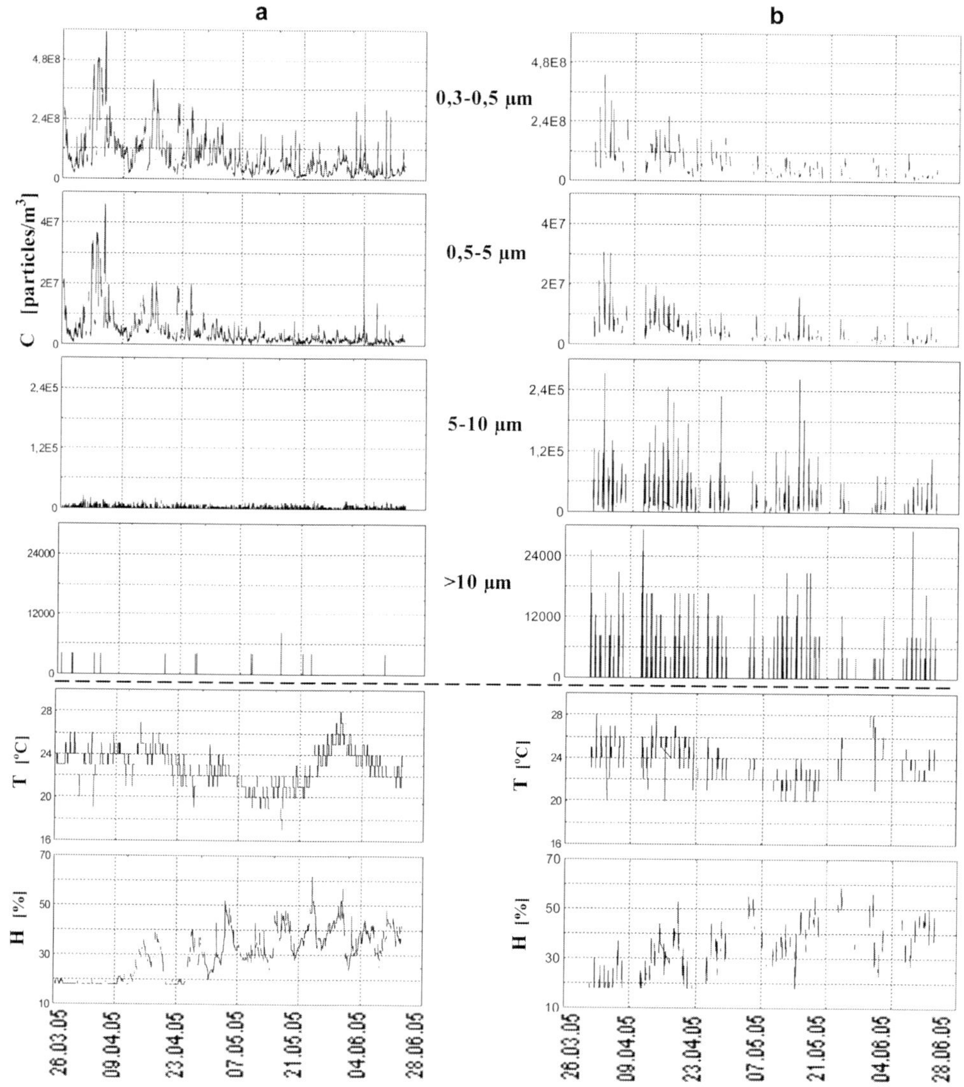

Figure 1. Changes of particle number concentration *C*, temperature *T* and relative humidity *H* in lecture room, a – without students, b – with students.

The results in Table 1 show that the average number concentration of coarse particles (5–10 and >10 μm) in the lecture room with students is clearly higher (t-test, $p < 0.001$) than in the lecture room without students. The ratio of extreme concentrations (Min and Max values) amounts to 0 and 14.5 for fraction 5–10 μm, 0 and 3.5 for fraction >10 μm. The ratio of average concentrations equals 20.7 and 287 for size fractions 5–10 μm and >10 μm, respectively. Furthermore, the average number concentration of the smallest measured particles (0.3–0.5 μm) in the lecture room with students is lower (t-test, $p < 0.001$) than in the lecture room without students.

The ratio of extreme concentrations amounts to 1.3 and 0.7, and the ratio of average values equals 0.9. Without including measurement errors, the results imply that the lecture room particle concentrations are affected by the presence and activities of students. This finding is consistent with the literature data (Hussein et al. 2005).

The relation between the concentration of aerosol particles, air temperature and relative humidity in a vacant lecture room and in an student-occupied room, is shown in Figure 2, in surface charts

Table 1. The variation of particle number concentrations (in particles m^{-3}) in lecture room.

Particle size, μm Particle number[a]	0.3–0.5 $\times 10^7$	0.5–5 $\times 10^6$	5–10 $\times 10^3$	>10 $\times 10^2$
without students				
Average	8.7	4.4	1.6	0.1
Median	6.2	2.4	1.5	0.1
SD	8.2	5.6	2.9	2.3
Min	0.7	0.4	0	0
Max	60.1	45.9	25.1	83.3
with students				
Average	7.8	5.2	33.1	28.7
Median	6.4	3.9	23.0	26.1
SD	5.9	4.1	35.7	42.8
Min	0.9	0.8	0	0
Max	43.0	30.8	364.0	292.4

[a]Average: arithmetic average; SD: standard deviation; Min: minimum; Max: maximum.

a and b respectively. It can be seen that concentration changes for small particles (size fractions 0.3–0.5 μm and 0.5–5 μm) have a minimum of around 30–40% relative humidity. The concentration increases for higher humidity with increasing air temperature, and for lower humidity with decreasing air temperature. To clarify these results, more detailed research which will also take into account fine and ultrafine particles must be done. It is to be assumed that processes which lead to a thermodynamic (thermal, mechanical and chemical) equilibrium state of aerosol systems, and such mechanisms as nucleation, coagulation, condensation and deposition (particle diffusion, thermophoresis, turbophoresis and gravitational settling) as well as condensation-evaporation processes which take place on particle surfaces, matter here. Nevertheless, the number concentration of small particles determined in changing indoor thermal conditions may have substantial consequences. The content of coarse particles (size fractions 5–10 μm and >10 μm) is indifferently correlated with lecture room air temperature and humidity, and is mainly influenced by the mobile activity of students (Fig. 2b).

The presented results indicate that the variation of particle concentrations in a lecture room is dependent on outdoor conditions and occupation as well as being highly affected by student activity. The results also show that the particle concentrations could be influenced by the thermal parameters of the indoor air.

The possible influence of the thermal parameters of the lecture room air on the concentration of aerosol particles is also shown in Figure 3. Distributions of indoor air specific enthalpy, for given concentrations of size selected particles, in the lecture room with and without students, are illustrated in diagrams a and b, respectively. The mean values of specific enthalpy (h_m) and the standard deviations (*SD*) are also shown in the diagrams. The specific enthalpy was determined for the integer values of air temperature and with the assumption that the measured air temperature was wet bulb temperature. To some extent, this allowed taking into account the influence of air humidity on the results of the particle measurements. From the distributions for all the selected sizes of particles, the values of h_m in the lecture room without and with the presence of students were about 67 kJ/kg and 72 kJ/kg and the *SD* were about 6.0 kJ/kg and 6.3 kJ/kg, respectively. It may indicate that the quantification of the considered particles in a given lecture room could be done using the results of the indoor air temperature and humidity measurements. Generally it could be expected that for every initially measured indoor environment, the concentration of such aerosol particles could be, with predictable error, determined on the basis of indoor air thermal parameters. Nonetheless, more detailed concessive research should be performed.

A comparison of the concentrations of all the measured indoor particle size fractions and the number of students (*SN*) present in the lecture room revealed that the concentration of small particles is poorly correlated with *SN* (0.3–0.5 μm, $r = -0.15$; 0.5–5 μm, $r = -0.10$). However, a slight improvement was noticed with a decrease in particle size. A better (positive) correlation

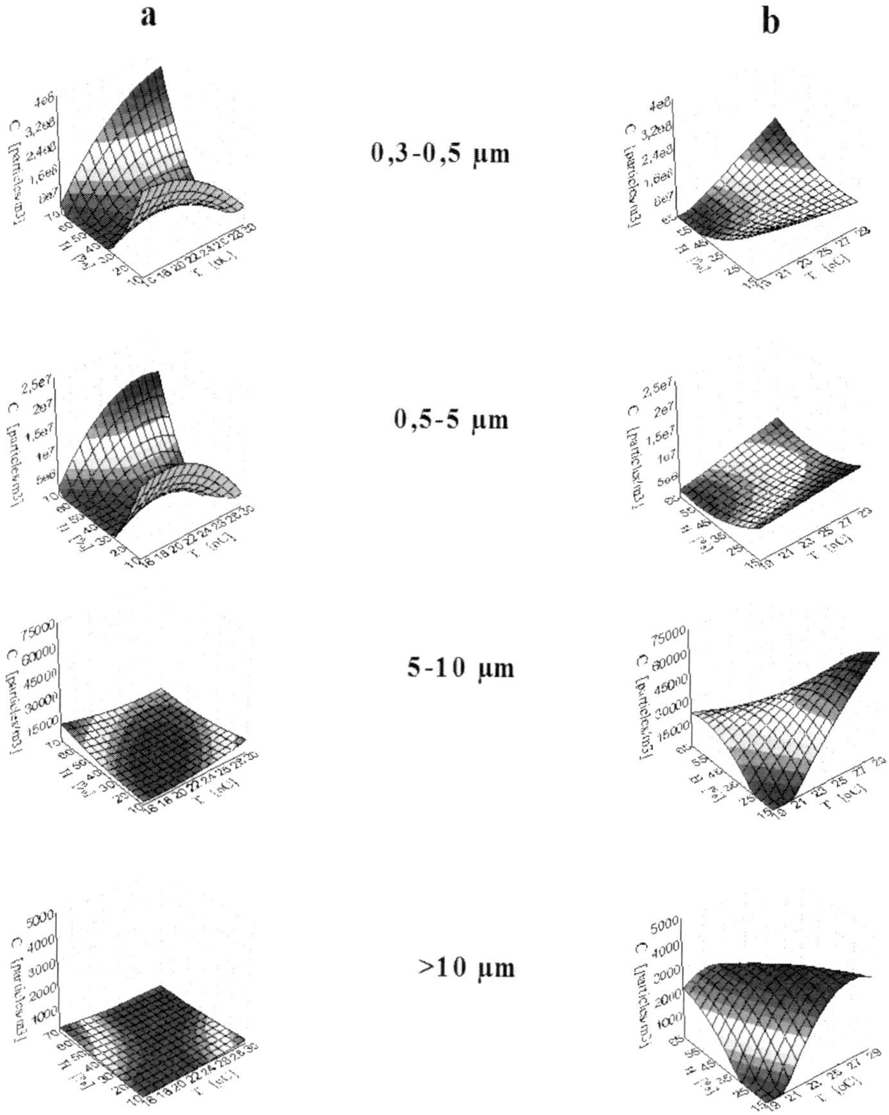

Figure 2. Relations between the indoor air temperature T, relative humidity H and particle number concentration C in the lecture room, a – without students, b – with students.

(but not significant $p < 0.05$) is observed between SN and the concentration of coarse particles (5–10 µm, $r = 0.32$; >10 µm, $r = 0.11$). This correlation coefficient for particles >10 µm deteriorated probably because of the different teaching procedures that were practiced in the lecture room, e.g. writing on the blackboard vs. on the whiteboard.

It is visible that the number concentration of the smallest particles measured (0.3–0.5 µm) decreased when students were present in the lecture room. On the other hand, the number concentration of coarse particles (>10 µm) increased very much. This reduction of small particles occurred in spite of the fact that the mean indoor air temperature and humidity had slightly risen (Figs 1, 4). According to the earlier shown dependences (Figs 2, 5), the increase of these particle number concentrations should be observed. One can speculate that this occurrence could be the result of the accumulation of certain amounts of small particles in students' lungs. Considering the fact that the sizes of most of the respirable particles were up to 5 µm (Sajo et al. 2002), such a

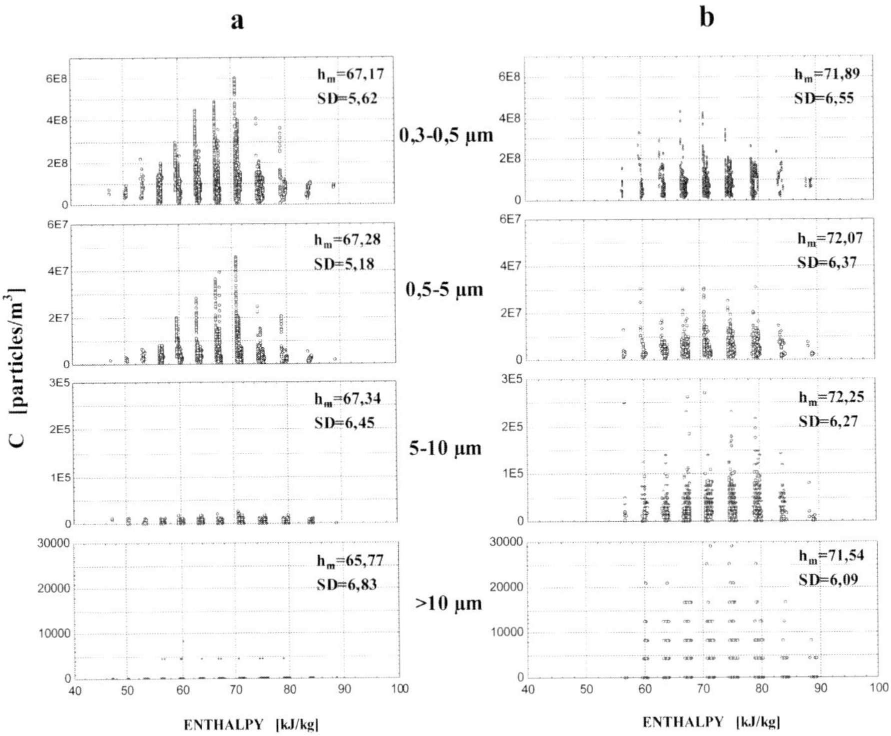

Figure 3. Distributions of specific enthalpy *h* in the lecture room, a – without students, b – with students.

thesis needs more detailed study. If the observed relations are thought to be causal, they could have consequences for student health because indoor particles inhaled and accumulated in their lungs are usually not neutral.

4 CONCLUSIONS

The indoor thermal conditions could affect the concentration of aerosol particles in a lecture room and consequently the number of accumulated particles in the lungs of the students. The concentration of small particles (<5 μm) decreases and that of coarse particles (>5 μm) increases when students are present in a lecture room. The mean values of air specific enthalpy distributions for the given indoor microclimatic conditions are the same for all determined particle size fractions, and depend on the occupancy of the lecture room. One of the solutions which could alleviate aerosol particle concentrations in the examined lecture room would be to improve room ventilation or to install air conditioning system. The proper formation of indoor air thermal parameters in a lecture room seems to be simple method to reduce the exposure of students to potentially harmful particles. The above general remarks should be treated as an introductory to this pilot study and have to be confirmed and developed in future research which will also include investigations of fine and ultrafine aerosol particles.

ACKNOWLEDGEMENTS

This research was carried out within the Projects 4955/T02/2008/34, financed by the Polish Ministry of Science and Higher Education.

REFERENCES

Afshari, A., Matson, U. & Ekberg, L.E. 2005. Characterization of indoor sources of fine and ultrafine particles: a study conducted in a full-scale chamber. *Indoor Air* 15(2): 141–150.

Asmi, A., Kulmala, M. & Pirjola, L. 1999. Modelling of indoor air aerosol size distributions with a sectional model. *J. Aerosol Sci.* 30: S719–S720.

Braniš, M., Řezáčová, P. & Domasová, M. 2005. The effect of outdoor air and indoor human activity on mass concentrations of PM_{10}, $PM_{2.5}$ and PM_1 in a classroom. *Environ. Research* 99: 143–149.

Franck, U., Herbarth, O., Wehner, B., Wiedensohler, A. & Manjarrez, M. 2003. How do the indoor size distributions of airborne submicron and ultrafine particles in the absence of significant indoor sources depend on outdoor distributions? *Indoor Air* 13(2): 174–181.

Gorbunov, B., Clarke, A.G. & Hamilton, R.S. 1999. Coagulation of soot particles and fractal dimension. *J. Aerosol Sci.* 30: 45–446.

Holmberg, S. & Chen, Q. 2003. Air flow and particle control with different ventilation systems in a classroom. *Indoor Air* 13(2): 200–204.

Hussein, T., Hämeri, K., Heikkinen, M.S.A. & Kulmala, M. 2005. Indoor and outdoor particle size characterization at a family house in Espoo-Finland. *Atmos. Environ.* 39: 3697–3709.

Jankowska, E., Kondej, D. & Jankowski, T. 2003. Particle size distribution and aerosol concentration at office workplaces. *J. Aerosol Sci., Abstracts of the European Aerosol Conference* 1: 391–392.

Kramer, I., Pöschl, U. & Niessner, R. 1999. Changes in the microstructure of aerosol particles after humidification. *J. Aerosol Sci.* 30: 307–308.

Lai, A.C.K. 2002. Particle deposition indoors: a review. *Indoor Air* 12(4): 211–214.

Morawska, L., He, C., Hitchins, J., Mengersen, K. & Gilbert, D. 2003. Characteristics of particle number and mass concentrations in residential houses in Brisbane, Australia. *Atmos. Environ.* 37: 4195–4203.

Nazaroff, W.W. 2004. Indoor particle dynamics. *Indoor Air* 14: 175–183.

Park, S.H., Lee, K.W., Shimada, M. & Okuyama, K. 2002. Change in particle size distribution of aerosol undergoing condensational growth: alternative analytical solution for the low Knudsen number regime. *J. Aerosol Sci.* 33: 1297–1307.

Połednik, B., Raczkowski, A. & Dumała, S. 2007. Aerosol particle concentration and the thermal conditions in a lecture room. *Environmental Engineering* – Pawłowski, Dudzińska, Pawłowski (eds), Taylor & Francis Group, London, 387–390.

Połednik, B., & Dudzińska, M.R. 2008. Occupation, thermal condition and aerosol particle number concentration in a lecture room. XII International Conference on Air Conditioning Air Protection & District Heating, Wroclaw, 439–444.

Połednik, B. & Dudzińska, M.R. (2010) Ventilation Control Based on the CO_2 and Aerosol Concentration and the perceived Air Quality Measurements – a Case Study. *Archives Environ. Prot.* 36(4): 67–80.

Pöschl, U. 2002. Formation and decomposition of hazardous chemical components contained in atmospheric aerosol particles. *J. Aerosol Med.* 15: 203–212.

Reff, A., Turpin, B.J., Porcja, R.J., Giovennetti, R., Cui, W., Weisel, C.P., Zhang, J., Kwon, J., Alimokhtari, S., Morandi, M., Stock, T., Maberti, S., Colome, S., Winer, A., Shendell, D., Jones, J. & Farrar, C. 2005. Functional group characterization of indoor, outdoor, and personal $PM_{2.5}$: results from RIOPA. *Indoor Air* 15(1): 53–61.

Risk assessment in relation to indoor air quality. 2000. Report No 22 EUR 19529EN.

Sajo, E., Zhu, H. & Courtney, J.C. 2002. Spatial distribution of indoor aerosol deposition under accidental release conditions. *Health Phys.* 83(6): 871–883.

Steinemann, A. 2004. Human exposure, health hazards and environmental regulations. *Environmental Impact Assessment Review* 24(7–8): 695–710.

Toivola, M., Nevalainen, A. & Alm, S. 2004. Personal exposures to particles and microbes in relation to microenvironmental concentrations. *Indoor Air* 14(5): 351–359.

Management of Indoor Air Quality – Dudzińska (ed)

Thermal modernization and biodeterioration of prefabricated elements of buildings – a case study

M. Piontek, M. Jasiewicz & K. Łuszczyńska
Institute of Environmental Engineering, University of Zielona Góra, Zielona Góra, Poland

ABSTRACT: Thermal modernization is carried out in a number of buildings. Wooden window frames are more and more often being replaced by plastic ones. A change in conditions and ventilation after the installation of new airtight windows can cause increased dampness inside a building, and mould. The tenants of infested buildings are exposed to allergens, volatile metabolites MVOC and the mycotoxins of moulds present in the partitions. In the Lubuskie Province the condition of the thermal protection in buildings built in the 1980s, was examined with a thermal camera. The results of this examination showed the presence of thermal bridges. After the windows in these buildings were replaced their air-tightness increased considerably. The result was the lack of air circulation in the flats and ventilating ducts. The damp, wet wall surfaces were infested with mould. A mycological analysis showed that apart from moulds such as Aspergillus vericolor, Penicillium chrysogenum, Scopulariopsis fusca, Acremonium charticola or Cladosporium herbarum, bacteria, yeasts, and mites were also present on partitions. Such infestations are a health danger to the inhabitants, especially as far as allergies are concerned.

Keywords: mycological examination, quality of indoor air, thermal modernization of buildings, thermal bridges, natural ventilation

1 INTRODUCTION

The fungi and bacteria concentrations in indoor air are closely connected with human health. At the end of the 1990s, Zyska (Zyska 1996) published data indicating that in Poland 8 million people in 2.7 million flats (out of a total of 11 million) were at risk from the allergens and mycotoxins of fungi present in building and interior finish materials. Six million people in 2 million flats were exposed to the effects of mould. The rotting of building timber is caused by mould. In recent years, in a number of scientific institutions all over the world, research has been carried out characterizing the bioaerosol of bacteria and mould in homes. A number of authors have proven that there is an important correlation between the level of indoor air pollution from fungal spores and particular illnesses of the tenants (Piotrowska et al. 2001).

The assumption of criteria for the assessment of mould deterioration of buildings is a difficult task. Attempts to specify standards for the quality of indoor air did not solve the problem. A submission made by Miller (1998), who associated an acceptable concentration of fungal spores in the air with the toxicity of species dominating in the room was rejected. He formulated his requirements as follows:

- the air in homes should not contain pathogenic and toxicogenic moulds such as *Aspergillus fumigatus* and *Stachybotrys chartarum*,
- the state of normal cleanness is maintained if the quantity of moulds present in the air is less than:
  - 150 cfu/m^3 for a mixture of various species of fungi,
  - 300 cfu/m^3 for species typical of the micro-flora of atmospheric air (Cladosporium, Alternaria).
- if the deterioration of air quality is indicated by the presence of moulds of a single species in a quantity greater than 50 cfu/m^3, further mycological tests are then required.

Table 1. Scale for the assessment of pollution in atmospheric air according to the standard PN-89/Z-04111/03.

Level of pollution in atmospheric air	Overall quantity of fungi (cfu/m^3)
Averagely clean atmospheric air, especially in late spring and early autumn	from 3000 to 5000
Pollution that may have a negative influence on man's natural environment	over 5000 to 10 000
Pollution posing a risk to man's natural environment	over 10 000

Table 2. Scale for the assessment of indoor air pollution according to Doležal (1990).

Description of conditions in a room	Acceptable quantity of fungi (cfu/m^3)
Very good mycological cleanness with multi-species micro-flora	100–300
General hygienic violation, indication for further tests	>500
Very high pollution, active process of moulding, domination of fungal species	10^5 and more

Table 3. D-A-N scale for the diagnostics of mould biodeterioration of buildings (Rymsza 2003).

Level of mycological contamination D-A-N	Description of the state of mycological contamination of the air or/and the surface of materials	Indoor air		Surface of building material	
		Quantity of fungi cfu/m^3	Quantity of ergosterol μg/m^3	Quantity of fungi cfu/100 cm^2	Quantity of ergosterol mg/m^2
Acceptable	normal state of pollution	<500	<0,01	<10^3	<2
Alarming	high state of pollution, without active growth of fungi	500–10^3	0.01–0.03	10^3–10^5	2–4
Dangerous	active growth of fungi	>10^3	>0.03	>10^5	>4

In Poland there are 5 laws concerning the assessment of atmospheric air:

- General directions and provisions about taking atmospheric air samples (PN-84/Z-04008/02, PN-89/Z-04008/08),
- Methods of microbiological examination of air (PN-89/Z-04111/01),
- Specifying the quantity of bacteria and microscopic fungi in atmospheric air when samples are taken by means of aspiration and sedimentation (PN-89/Z-04111/02, PN-89/Z-04111/03), (Table 1).

As far as indoor air in tenement buildings is concerned, these acceptable quantities of mould were suggested by Krzysztofik (1992):

- bedroom <100 cfu/m^3,
- rooms <200 cfu/m^3,
- kitchen <300 cfu/m^3.

The scale for the assessment of indoor microbiological cleanness according to Doležal (Doležal et al. 1990) is presented in Table 2. Rymsza (2003) suggested a scale for the diagnostics of mould bio-deterioration of buildings, with the use of the D-A-N scale (Table 3).

In 2004, a Team of Experts on Biological Factors of the Inter-Department Committee for the Highest Acceptable Concentrations and Intensities of Agents Harmful to Health at Work Stations suggested the adoption of recommended values of acceptable concentrations of the most common categories of microorganisms and bacterium endotoxin in indoor air (Table 4). These recommendations, which have been used in Poland for a couple of years are helpful in the assessment of the

Table 4. Suggestions for recommended concentrations of microorganisms and endotoxins in indoor air prepared by the Team of Experts on Biological Factors (Górny 2009).

	Acceptable concentration	
Microbiological factor	Work station rooms polluted with organic dust	Living and public quarters
Mesophilous bacteria	1.0×10^5 cfu/m³*	5.0×10^3 cfu/m³
Gram-negative bacteria	2.0×10^4 cfu/m³*	2.0×10^2 cfu/m³
Thermophilous actinomycetes	2.0×10^4 cfu/m³*	2.0×10^2 cfu/m³
Fungi	5.0×10^5 cfu/m³*	5.0×10^3 cfu/m³
Factors causing serious diseases in humans	0 cfu/m³	0 cfu/m³
Bacterium endotoxin	200 ng/m³ (2000 EU/m³)**	5 ng/m³ (50 EU/m³)

* for the respirable fraction the suggested values should be lower by half and amount to: 5.0 × 104 cfu/m³ for mesophilous bacteria; 1.0 × 104 cfu/m³ for Gram-negative bacteria; 1.0 × 104 cfu/m³ for thermophilous actinomycetes; 2.5 × 104 cfu/m³ for fungi and 100 ng/m³ (1000 EU/m³) for bacterium endotoxin
** EU – Endotoxin Units

exposure to biological factors at work stations or indoors. The recommendations are also helpful for adopting the right precautions and preventive measures to be used in these environments (Górny 2009).

Other authors maintain that in epidemiological research, a better measure for the determination of mould exposure is an examination of the biomasses of moulds rather than an examination of cfu/m³ of air. Concern for the sanitary-hygienic conditions in buildings is included in the directive of the Ministry of the Interior and Administration from 1999: "in the form of technical conditions that buildings and their location should satisfy" (Journal of Bills 1999 No. 15 item 140). This directive says that the construction and choice of materials for external partitions, conditions for temperature and dampness, as well as the intensity of air circulation in rooms should prevent the appearance of moulds (Janińska 2000).

People who live flats need oxygen. The lack of oxygen can be felt after a couple of minutes. The amount of outdoor air coming into a room where people live or stay should only temporarily be 20 m³/h for each person.

The task for the ventilation system is to route fresh outdoor air into rooms in a planned and organized way, and then allow the air to flow through the rooms leaving the building as dead air through ventilating ducts, according to the standard PN-83/B-03430/Az3:2000. The efficiency of a ventilation system depends on:

- the right conveyance of fresh air into the building from the outside,
- ensuring the right circulation between rooms,
- efficient removal of dead air.

Neglecting any of the three points causes disturbances in ventilation or even makes the ventilation of rooms impossible.

The driving force of natural ventilation is the difference between the density of cold air (fresh air outside) and warm air (air in rooms). Cold air flows into rooms through gaps in the windows and walls of a building or through intake ventilators installed in the windows and walls. After being mixed with the air inside and warmed up, it leaves the room – together with pollution, through air grates connected to the ventilating ducts. According to the Polish building code inlets into ventilating ducts should be situated in the kitchen, in the bathroom and in any separate room with a toilet. The minimum amount of air circulation should be:

- 70 m³/h for a kitchen with an outside window and a gas cooker,
- 30 m³/h for a kitchen with an outside window and an electric cooker in a flat for three people, and the same for a separate room with a toilet,
- 50 m³/h for more than three people and for a bathroom,
- 20 m³/h for one person in a flat (30 m³ if the person smokes).

Airtight windows do not ensure the right supply of air for ventilation. As a result, the chimney draught is weakened, air circulation is reduced, and pollution accumulates inside the building. Bad ventilation in a building results in problems such as: dampness and microbiological pollution.

The goal of this paper is to prove that a partial thermal modernization renovation does not bring desirable results. A short time after such a renovation, mould appears on the walls of the flats. Thermal modernization in buildings should make sure the air-flow in rooms is the same as before the renovation. Air-flow should be in agreement with building standards, both before and after renovation. After rooms are insulated with the help of the installation of new windows, dampness increases in the rooms and thermal irregularities appear on the surfaces of the building.

Both phenomena have been proved in this paper by the results of mycological and thermographic analyses. Our research showed that the main cause of biodeterioration was improper ventilation after the installation of new airtight windows

In order to assess the state of the thermal protection of the buildings, thermovision examinations were carried out. The main purposes of the examinations were, 1) to identify thermal bridges and 2) to assess the quality of the materials used for thermal homogeneity of the partitions in question. A thermovision device makes it possible to visualize the field of the surface temperature in the form of a so-called thermogram. Different properties of the thermal insulation layer in the partition produce a varied layout of temperatures on the surface of the partition. In turn, infrared ray emissions take on a varied character. This energy is transformed by an infrared camera into electrical impulses which form a picture of the surface under research. It is then possible, to draw conclusions about the thermal insulation quality of the partition (Krause 2008).

2 MATERIALS AND METHODS

A mycological analysis, thermographic measurements, and measurements of dampness were carried out in a 25 year old building built of large prefabricated elements. Four flats were examined where mould could be seen on the walls. The research was carried out on the following rooms:

- ground floor flat No. 1: the kitchen and the bathroom,
- ground floor flat No. 2: the kitchen and the living room,
- first floor flat No. 3: the kitchen,
- second floor flat No. 4: a room.

2.1 *Thermographic measurments*

The measurements were carried out with a thermographic camera; series TiR2FT-0903638, version 3.0.7.3 produced by the company Fluke. Thermal sensitivity of the camera was from 0.07°C.

The temperature of indoor and outdoor air was measured using a mercury thermometer placed 2 meters from the object under research.

General rules about thermovision measurements in buildings specified by the standard PN EN 13187: 2001 were followed.

2.2 *Mycological analysis*

The mycological analysis was carried out according to the methodology specified by CBS (Centraalbureau voor Schimmelcultures), (Samson et al. 2004). From among the suggested methods for detecting and isolating different kinds of moulds in a house, the method of direct inoculation into Petri dishes was applied. Tiny bits of material infested with fungi are placed or sprinkled on the breeding base. (Hoekstra et al. 2004). Direct inoculation into dishes is regarded as one of the more effective techniques of mycological analysis of materials.

The material was inoculated directly in the place where the mould appeared on the surface of four nutrient media (two synthetic nutrient media, Czapek-Dox Agar (Cz), Synthetic Nutrient Agar (SNA) and two natural nutrient media Malt Extract Agar (MEA), Potato Dextrose Agar (PDA) on Petri dishes. The material was sprinkled on the breeding base in a small quantity with a small sharp spade-shaped tool. The samples were covered with white canvas and incubated in a breeding room at a temperature of 18°–22°C, The day and night rhythm was preserved. Pure (axenic) cultures

Table 5. Results of the mycological and thermal analyses in the staircase.

Lp.	Species of micro-organisms in the air and on the partitions	Quantity of mould fungi in the air (cfu/m^3) and on the partitions (cfu/100 cm^2)	Dampness of the partitions (%)	Temperature of the air and the partitions (°C)
Air of the staircase not infested with mould				
1.	*Penicillium chrysogenum* *Cladosporium herbarum* *Rhizopus nigricans* *Mucor racemosus* *Penicillium expansum* Bacteria	*2.1 × 10^2	–	Air 21.0
Partitions in the staircase without not infested with mould				
2.	*Penicillium chrysogenum* *Cladosporium herbarum* *Mucor racemosus* Bacteria yeasts	**5.1 × 10^2	2.5 % – dry wall	Partition 18.1

*sediment method
**contact method

were isolated from the initial mixed cultures and passaged onto two nutrient media: Czapek-Doxa (Cz) and malt agar (MEA). The time of inoculation, breeding, and observation for an isolated species was approximately 21 days. It is the total amount of time needed for breeding and isolating species from mixed samples taken from the walls (about 5–10 days) to produce pure species (about 5–10 days). Breeding practice showed that approximately 21 days were needed to obtain colourful sporing cultures (Piontek 2007, 2010). The stains of moulds obtained in this way were then examined to identify their taxonomic membership. Mould which belongs to a particular species was determined in a particular species was determined macroscopically and microscopically on the basis of morphological and physiological features in agreement with the methodology specified (Samson et al. 2004). In order to determine the quantity of organisms in the colonies of microorganisms in the dishes, the visual assessment method was used.

A visual scale was used to assess the colonies in terms of quantity: very numerous, numerous, quite frequently, singly (authors' own unpublished research). The colonies in the Petri dishes were observed macroscopically with a binocular (stereoscopic microscope). The moulds in the dishes and the moulds after preparations had been made were observed microscopically with a Nikon TMS microscope.

However, the scale: abundant, moderate, single was used in order to assess the extent of wall infestation.

2.3 *Sediment method*

In order to compare the results from mould infested flats, measurements were also taken in the staircase which was not infested with mould. The analysis of the air was carried out by means of the sediment method on the MEA and Cz nutrients. In order to achieve this, open dishes with the nutrient were left out for about 15 minutes (Hoekstra et al. 2004). In addition, measurements were taken on partitions by means of the contact method. The results are presented in Table 5.

2.4 *Measurements of wall dampness*

The measurements of the dampness of the partitions infested with moulds, were carried out with a Trotec T650 hygrometer and assessed according to the scale included in the manual of the device (Table 6). Absolute dampness (%) was measured at the depth of 2–4 cm, depending on the density related to the total volume of the material. The instrument is an electronically calibrated volumetric

Table 6. Efficiency of ventilating ducts expressed by the temperature and air flow velocity.

	Measurements of the temperature and the air flow velocity	
Object under examination:	Air temperature (°C)	Air flow velocity (m^3/h)
Object No. 1:		
Kitchen	23.4	33.3
Bathroom	24.0	26.8
Object No. 2:		
Kitchen	21.0	18.7
Bathroom	22.0	0.0
Object No. 3:		
Kitchen	23.3	21.2
Bathroom	24.1	8.6
Object No. 4:		
Kitchen	24.5	37.3
Bathroom	25.6	35.3

Table 7. Assessment of dampness of the wall.

Mass dampness of the wall %	Assessment of the state of dampness of the wall
0–3	dry wall
3–5	wall with more than average dampness
5–8	wall with medium dampness
8–12	wall with high dampness
>12	damp walls

dampness meter which can be used to measure dampness or localise damp areas quickly and non-invasively.

2.5 *Measurements of air velocity at the ventilating duct outlets*

The air velocity and temperature at the outlets of the ventilating ducts in kitchens and bathrooms were measured using a portable anemometer Testo 425, with a permanently attached flow probe (the diameter of the gauging point – 7.5 mm). The meter had a telescopic hold (max. 820 mm), a battery, and a calibration protocol. The measurement range of the anemometer was: $0 \ldots +20$ m/s, accuracy: $\pm(0.03$ m/s $+ 5\%$ of the measurement value), resolution: 0.01 m/s. The measurements were taken in kitchens and bathrooms of the objects under research, with the windows and doors tightly closed. The results were recalculated in m^3/h units and are presented in Table 7.

3 RESULTS AND DISCUSSION

The thermographic, mycological and dampness examinations were carried out in late autumn, in November 2009. The thermographic measurements were first done outside the building, and then inside in the places where mould appeared on partitions. The temperature was measured outside and inside, and the dampness of the walls was measured.

3.1 *Measurements outside the building*

Measurements outside the building are illustrated in Figure 1 and Figure 2. Parameters measured outside the building are presented in Table 8. The measurements were taken on the 25th of November,

Table 8. Parameters measured outside the building.

Date of measurements	25.11.2009 r.
Time of measurements	3 p.m.–4 p.m.
Temperature of external air	13°C
Relative dampness of the external partition	8.1% (very damp wall)
Wind	no
Rain	no
Technical description of the object under research	Location of the block under research: Sława, Poland Age of the building: 25 years Materials of which the building was built: prefabricated elements Facade: plaster Windows: plastic windows in the whole block Ventilation: lack of an operational ventilation system Heating: central heating Causes of mould infestation: thermal bridge in all the objects under research

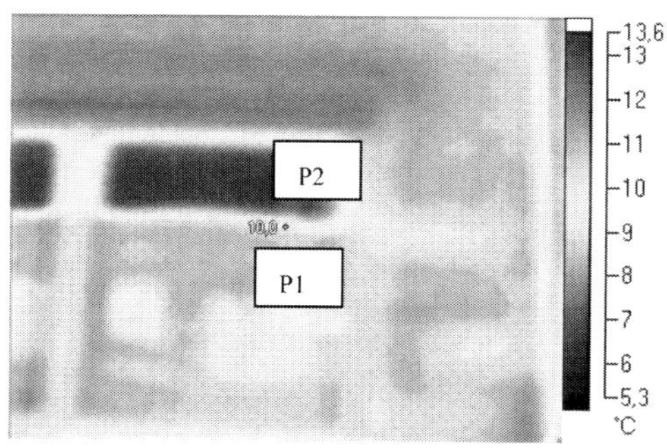

Figure 1. Thermographic picture of the building facade; P1 – 6.3°C; P2 – 11.8°C.

2009 between 3 p.m. and 4 p.m. It was a cloudy day without rain or wind. Junctions of the elements can be seen where the temperature on the surface of the wall is higher, because these places are practically devoid of any thermal insulation.

The minimum temperature of the surface of the external partition was 6.3°C (point P1), maximum 11.8°C (point P2). In an infrared photo (Figure 1), the place where inside heat is escaping can be clearly seen. A histogram presenting the layout of temperatures is shown in Figure 3. The infrared photos show that the building is insufficiently insulated. Thermal protection of the building is advisable. The infrared photos show a number of thermal bridges where heat is escaping from heated rooms.

3.2 *Results of thermographic measurements from four flats infested with mould*

Flat No. 1 – kitchen

The photo shows the external walls of a kitchen (Figure 5). The corner between the walls is infested with mould, the area with mould extends between the upper corner of the wall and the floor. Because of an intense flow of air outside in the corners of the building, this corner has the lowest temperature – 16.1°C. Because of a relatively high quantity of dampness in the rooms, dew may appear on the walls. The temperature of the walls with dew will be low if the temperature falls below the dew point for air inside the room.

The infrared photo (Figure 4) shows that the values of the temperatures in that area are as follows:

- maximum: 18.3°C (point P2),
- minimum: 16.1°C (point P1).

Figure 2. Facade within the visual range.

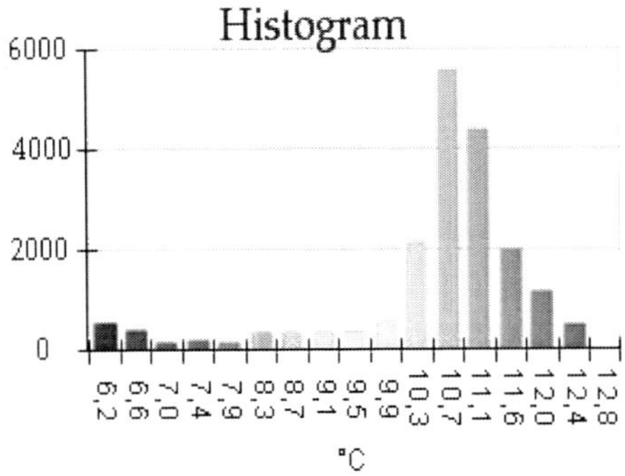

Figure 3. Histogram of the layout of temperatures.

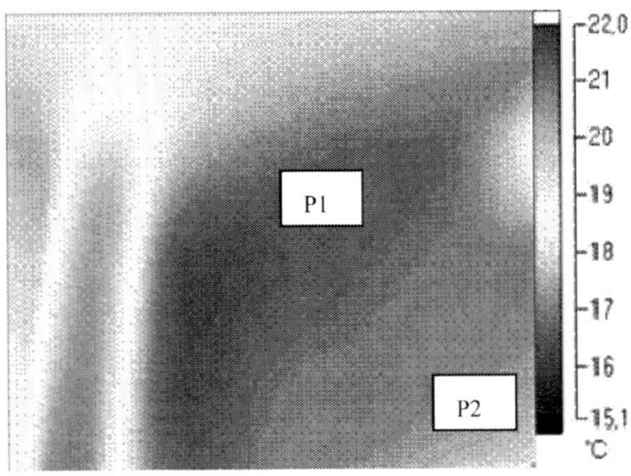

Figure 4. Thermal bridge on the partitions in room No. 1 (kitchen) P1 – 16.1°C; P2 – 18.3°C.

The area most infested with mould has the lowest temperature, which is shown in blue in the infrared photo. The maximum temperature is the temperature of the central heating pipe – red. A thermal bridge in the kitchen of flat No. 1 seen in a 3D picture is illustrated in Figure 6.

Flat No. 1 – bathroom

The infrared photo (Figure 7) shows the lower corner of the window in the bathroom. Mould can be seen on the wall adjacent to the window. The mould is in the lower corner of the wall next to the window. The junction between the inside wall and the front wall within the window opening. The space between the inside wall and the front wall where the window is positioned is the cause for the appearance of thermal bridges. A part of the window seen from the inside of the building is presented here, in which the results of the existence of a thermal bridge can be seen. The value of

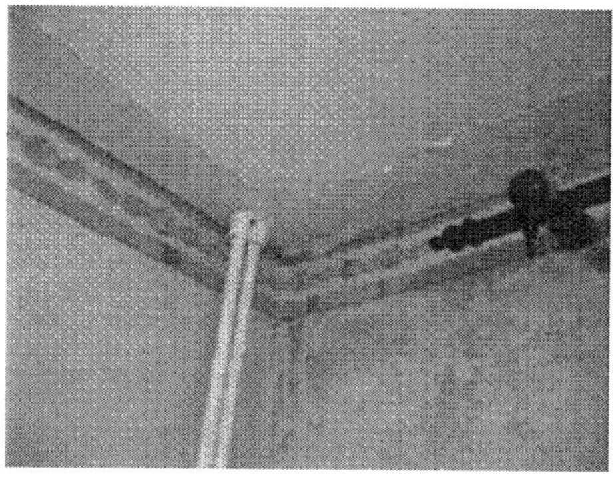

Figure 5. Corner of the kitchen (room No. 1).

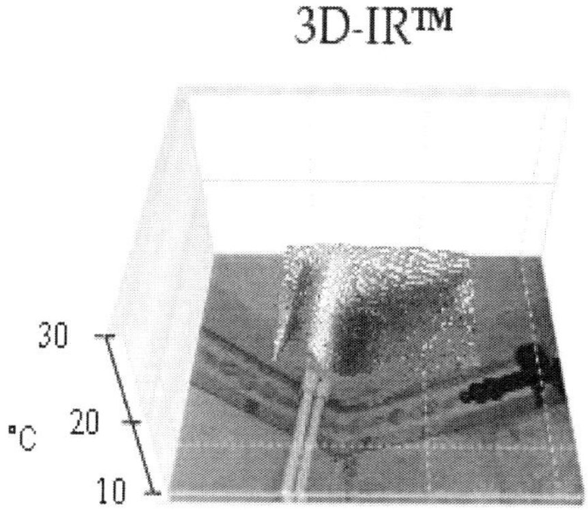

Figure 6. Thermal bridge in the kitchen of flat No. 1 seen in a 3D picture (background temperature 20.0°C, emissivity 0.95).

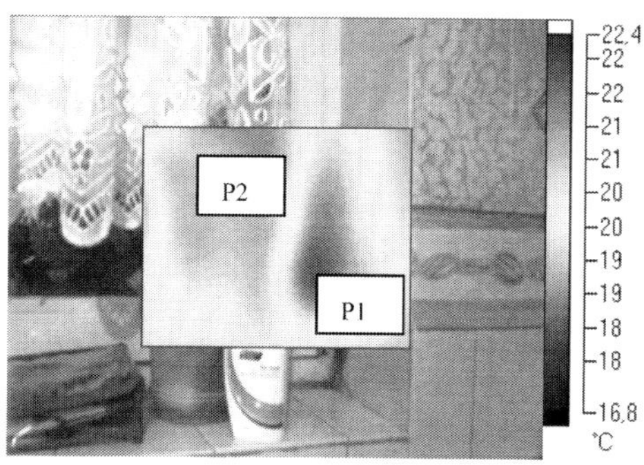

Figure 7. Thermal bridge on the partitions in flat No. 1 (bathroom) P1 – 17.8°C, P2 – 21.4°C, (background temperature 20.0°C; emissivity 0.95; average temperature 19.9°C).

the temperature given in the description of the figure – 17.8°C is the lowest temperature observed in that area. Such a low temperature value means that the wall has a low thermal resistance near the edge of the window opening. The low temperature of that part of the wall may be the cause for the appearance of dew on the surface of the wall and consequently, the appearance of mould.

The lowest temperature is 17.8°C (point P1), and the highest is 21.4°C (point P2). The place where heat is escaping from the room is marked in blue – a clearly seen thermal bridge – point P1, which indicates poor thermal insulation.

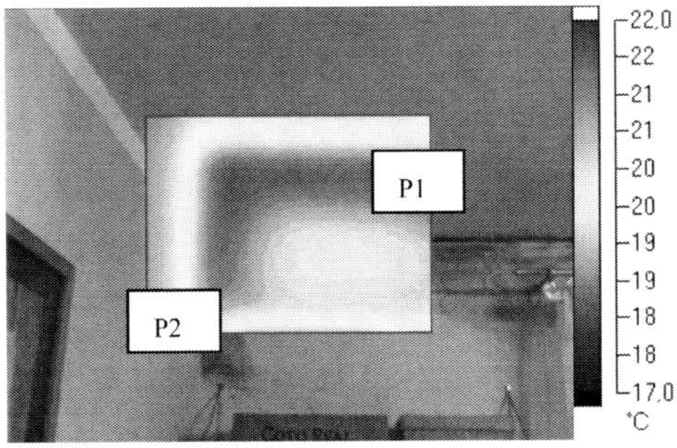

Figure 8. Thermal bridge – flat No. 2 (kitchen)) P1 – 18.0°C, P2 – 21.0°C, (background temperature 20.0°C; emissivity 0.95; average temperature 18.8°C).

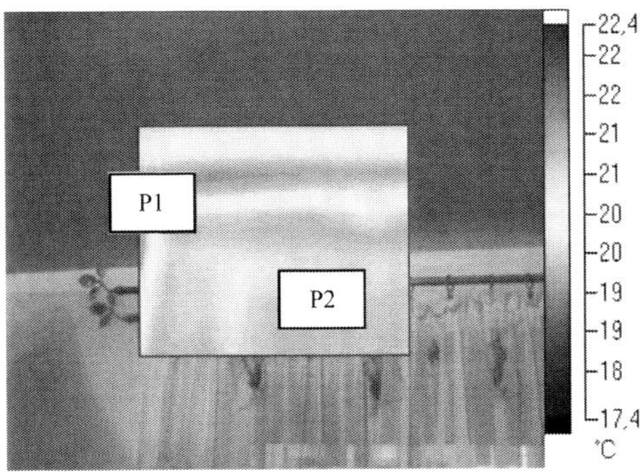

Figure 9. Thermal bridge in the room (flat No. 2) P1 – 18.8°C, P2 – 21.0°C, (background temperature 20.0°C; emissivity 0.95; average temperature 20.3°C).

Flat No. 2 – kitchen
The photo (Figure 8) shows the north corner of the external wall of the kitchen. A damp lintel can be clearly seen. There is a layer of mould on it extending from the upper corner of the kitchen to the floor. The temperature registered in the dampest place is 18.0°C (point P1). The maximum temperature is 21.0°C (point P2). The area most infested with fungi has the lowest temperature, which is marked in blue in the infrared photo. The thermogram clearly shows the presence of a thermal bridge.

Flat No. 2 – room
The infrared photo (Figure 9) shows that temperature values in that area are as follows:

- maximum: 21°C (point P2),
- minimum: 18.8°C (point P1).

The area the most infested with mould has the lowest temperature, which is marked in blue in the infrared photo. The thermogram clearly shows a thermal bridge which appeared as a result of bad insulation between the ceiling and the wall.

Flat No. 2 – kitchen
The infrared photo (Figure 10) shows that the values of the temperatures in that area are as follows:

- maximum: 20.6°C (point P2),
- minimum: 17.9°C (point P1).

The area the most infested with mould has the lowest temperature, which is marked in blue in the infrared photo. The thermogram clearly shows the presence of a thermal bridge.

Figure 10. Thermal bridge on the partition of ground floor flat No 3 – kitchen, P1 – 17.9°C, P2 – 20.6°C, (background Temperature 20.0°C; Emissivity 0.95; Average temperature 18.7°C).

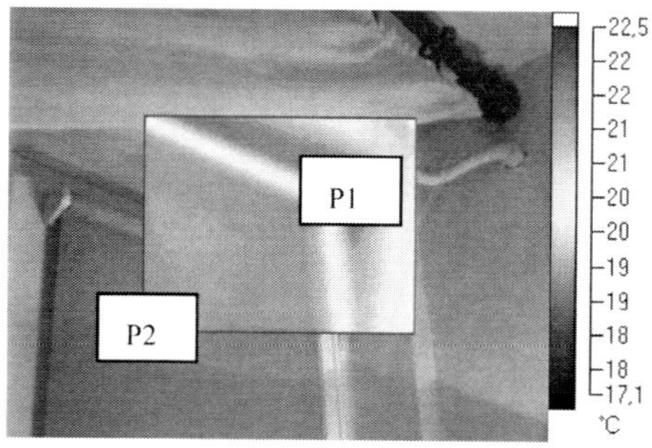

Figure 11. Thermal bridge in flat No. 4 – room P1 – 17.9°C, P2 – 20.06°C, (background temperature 20.0°C; emissivity 0.95; average temperature 18.7°C).

Flat No. 4 – room
The infrared photo (Figure 11) shows that the values of the temperatures in that area are as follows:

- maximum: 21.1°C (point P2),
- minimum: 17.4°C (point P1).

The area most infested with mould has the lowest temperature, which is marked in blue in the infrared photo. Cold parts of the walls can be seen in blue. The thermogram clearly shows the presence of a thermal bridge.

3.3 *Results of the measurements of dampness and temperature partitions*

In the objects under examination, the temperature and dampness of partitions were measured in the places infested with fungi. The results are presented in Table 9.

3.4 *Results of mycological analysis*

The mycological analysis showed that the most numerous types of moulds were: *Penicillium chrysogenum, Cladosporium herbarum*, and *Aspergillus versicolor*. In addition there were bacteria, actinomycetes and yeasts. In one of the flats, among the various moulds, there were *Acari mites* (Table 10).

The environment in rooms can include numerous potentially harmful substances such as: allergens, mites, formaldehyde, ozone and volatile fumes of organic substances. In current research, the greatest attention is paid to the presence, growth, and the prediction of the growth of fungi including mycotoxicogenic ones such as Aspergillus vericolor and Stachybotrys chartarum (Dales et al. 1991, Hendry et al. 1993, Johanning et al. 1996). For this reason mycological examinations were carried out as well as examinations from the field of building physics, in flats infested with fungi.

Table 9. Assessment of dampness of partitions in the objects under examination and the temperature of the partition.

Object under examination:	Measurement of dampness (%) and temperature of the partition		Assessment of the dampness of the wall
	Temperature of the partition (°C)	Dampness (%)	
Object No. 1:			
Kitchen	22.5	5.4	fairly damp wall
Bathroom	23.0	5.1	fairly damp wall
Object No. 2:			
Kitchen	17.4	9.3	very damp wall
Room	20.4	8.4	very damp wall
Object No. 3:			
Kitchen	23.3	5.1	fairly damp wall
Object No. 4:			
Living room	25.1	8.1	very damp wall

Badly insulated buildings, flats with above average dampness, and also flats clearly infested with mould can often be found despite the fact that all the rules of thermal insulation have been formulated and learned. The results of the research have proven that the cause of fungal biodeterioration is the presence of thermal bridges on partitions, and the consequences of the changes in the condition of those places i.e. insufficient ventilation for the air tightness of new windows. The thermographic measurements assessed the state of thermal protection of the buildings as incorrect. The following problems were found:

- incorrect thermal insulation at the junction of the external walls,
- natural ventilation not matched for the air tightness of new windows and decreased air circulation,
- less intensive airing, and lowereing the temperature inside while sleeping or when not at home, in the tenants' attempt to reduce heating costs.

Measurements of dampness in places infested with moulds showed high and medium dampness of the partitions.

Lowering the internal temperatures of the surfaces of the external partitions caused dampness to consolidate in the places where thermal bridges appeared. Apart from making the conditions of the thermal environment worse, it caused the biodegradation of building and finish materials. These results are shown by the thermograms presented in Fig. 1, 4, 7, 8, 9, 10, 11 and discussed in detail in the chapter "results". We compared the extent of mould infestation in the flats under study. The amount of mould depended on the behaviour of the tenants. The greatest mould infestation was observed in the ground-floor flat (flat No. 2) where a family with a small child lived. The tenants of that flat did not air the rooms. Laundry was hung to dry in the room which increased the dampness in the flat.

Different types of moulds and other microorganisms such as: bacteria, actinomycetes and mites appeared in the flats, which lowered their sanitary standard. Among the species of moulds isolated (Table 10) there were moulds potentially allergenic and toxicogenic such as: C. herbarum, P. chrysogenum and A. versicolor. The most numerous species were: P. chrysogenum, A. versicolor, C. herbarum, which are the most common in flats. A.versicolor frequently occurs in buildings, and this mould synthesizes carcinogenic sterigmatocystin (ST). For this reason, A.versicolor poses the greatest mycotoxical danger to people inhabiting buildings infested with mould (Piontek 2004). The risk of inhaling mycotoxins can appear when volatile particles such as dust or mouldy substances containing mycotoxins are inhaled. Particles of dust which have an aerodynamic diameter of $<7\ \mu m$ are regarded as "inhalable", larger particles are labelled "non-inhalable" ($>7\ \mu m$). The spores of a number of species of Penicillium and Aspergillus easily accumulate in airways because their size is Ø 2–5 μm.

Table 10. Biodeteriogenic organisms present in thermal bridges.

Species/kind	Quantity
flat No. 1/kitchen	
Penicillium chrysogenum Thom	abundant
Penicillium aurantiogriseum Diercks	moderate
Aspergillus versicolor (Vuillemin) Tiraboschi	single
Cladosporium herbarum (Persoon) Link ex S.F. Gray	moderate
Actinomycetes	single
Bacteria	single
flat No. 1/bathroom	
Cladosporium herbarum (Persoon) Link ex S.F. Gray	abundant
Aspergillus versicolor (Vuillemin) Tiraboschi	single
Penicillium chrysogenum Thom	single
Moniliella acetobutens Stolk & Dakin	single
Scopulariopsis fusca Zach	moderate
Actinomycetes	moderate
Bacteria	abundant
flat No. 2/kitchen	
Penicillium chrysogenum Thom	abundant
Cladosporium herbarum (Persoon) Link ex S.F. Gray	abundant
Aspergillus versicolor (Vuillemin) Tiraboschi	moderate
yeasts	single
Bacteria	single
flat No. 2/room	
Cladosporium herbarum (Persoon) Link ex S.F. Gray	abundant
Penicillium chrysogenum Thom	moderate
Aspergillus versicolor (Vuillemin) Tiraboschi	single
Acremonium chatricola (Lindau) W. Gams	single
Actinomycetes	moderate
Bacteria	abundant
Acari mites	moderate
flat No. 3/kitchen	
Cladosporium herbarum (Persoon) Link ex S.F. Gray	abundant
Aspergillus versicolor (Vuillemin) Tiraboschi	single
Penicillium chrysogenum Thom	single
Bacteria	single
flat No. 4/room	
Penicillium chrysogenum Thom	moderate
Cladosporium herbarum (Persoon) Link ex S.F. Gray	single
Bacteria	moderate
Actinomycetes	moderate

4 CONCLUSIONS

In developed countries there is a noticeable tendency for people to spend a lot of time in closed spaces. It is believed that the quality of the air in closed spaces has a considerable influence on human health.

The above research confirmed the cause and effect reaction between technological defects in buildings (thermal bridges), thermal modernization (changing windows), and disturbances in the ventilation system. All of these factors consequently lead to the biodeterioration of technological materials and the lowering of the sanitary condition of indoor air in buildings. Therefore, renovations directed at thermal modernization should be carried out on a broad basis, including all solutions to heat/dampness conditions, in order to prevent biodeterioration in buildings.

REFERENCES

Dales R.E., Zwanenburg H., Burnett R., Franklin C.A. 1991. Respiratory health effects of home dampness and moulds among Canadian children. *Am. J. Epidemiology* 134: 196–206.

Doleżal M., Doleżal M., Pieniążek Z. 1990. *Toxigenic fungi in leaving premises (in Polish: Grzyby pleśniowe w budynkach)* Łódź, SOS-PGM Inwestprojekt.

Flannigan B. 1997. Air sampling for fungi in indoor environments. *J. Aerosol Sci*. Vol. 28 (3): 381–392.

Górny R. 2009. Areozole biologiczne – rola normatywów higienicznych w ochronie środowiska i zdrowia. W: *V Konferencja Naukowa Rozkład i korozja mikrobiologiczna materiałów technicznych*, Politechnika Łódzka, Łódź, 91–102.

Hendry K.M., Cole E.C. 1993. A review of mycotoxins in indoor air. *J. Toxicol. Environ. Health* 38: 138–198.

Hoekstra E.S., Samson R.A., Summerbell R.C. 2004. Methods for the detection and isolation of fungi in the indoor environments. In: R.A. Samson, E.S. Hoekstra, J.C. Frisvad, (eds.) *Introduction to food and airborne fungi*. Seventh Ed. Utrecht. Centralbureau voor Schimmercultures (CBS) The Netherlands.

Janińska B. 2000. Czynniki środowiskowe determinujące aktywność korozji mikrobiologicznej w budownictwie. In: *Ogólnokrajowa konferencja Naukowa, Rozkład i korozja mikrobiologiczna materiałów technicznych*, Łódź, Politechnika Łódzka, 12–16.

Johanning E., Biagini R., Hull D., Morey P., Jarvis B., Landsbergis P. 1996. Health and immunology study following exposure to toxigenic fungi (*Stachybotrys chartarum*) in water-damaged office environment. *Int. Arch. Occup. Environ. Heath*. 68: 207–218.

Krause P. 2008. Stan ochrony cieplnej ścian zewnętrznych budynków ocieplonych styroblokami, Energia i budynek 7–8 (17) 2008, 26–29.

Krzysztofik B., 1992. *Mikrobiologia powietrza*. Warszawa. Wydawnictwo Politechniki Warszawskiej.

Miller J.D., Laflamme A.M., Sobol Y., Lafontaine P., Greenhalgh R. 1998. Fungi and Fungal Products in some Canadian Houses, *International Biodeterioration*, 24: 103–120.

Piotrowska M., Żakowska Z., Gliścińska A., Bogusławska-Kozłowska J. 2001. Impact of outdoor Nioberosols on indoor fungi In: Microbiological deterioration of buildings and technological materials (in Polish: Rola mikroflory powietrza zewnętrznego w kształtowaniu bioareozolu grzybowego pomieszczeń zamkniętych *W: Rozkład i korozja mikrobiologiczna materiałów technicznych, II Konferencja Naukowa)*, Łódź, 113–118.

Piontek M. 2007. Strains of Aspergillus versicolor Tiraboschi synthesizing sterigmatocistin and the differentiation of mycotoxic risk dependent on their productivity in housing buildings. *Mycotoxin Research* 23(1): 34–38.

Piontek M. 2010. Use of planarian *Dugesia tigrina* Girard bioassay for assessing the toxicity of sterigmatocistin produced by *Aspergillus versicolor* Tiraboschi. Environment Protection Engineering 36(1): 65–71.

PN-83/B-03430/Az3:2000. Polish Standard: Ventilation in public buildings and leaving premises – requirements.

PN-84/Z-04008/02. Polish Standard: Air Protection. Sampling – general rules. Warsaw, Polish Standardization Committee.

PN-89/Z-04008/08. Polish Standard: Air Protection. Air sampling for microbiological measurements – aspiration and sedimentation methods. Warsaw, Polish Standardization Committee.

PN-89/Z-04111/01. Polish Standard: Air protection. Microbiological measurements – general rules. Warsaw, Polish Standardization Committee.

PN-89/Z-04111/02. Polish Standard: Air Protection. Number of bacteria measurements by aspiration and sedimentation methods. Warsaw, Polish Standardization Committee.

PN-89/Z-04111/03. Polish Standard: Air protection. Number of fungi measurements by aspiration and sedimentation methods. Warsaw, Polish Standardization Committee.

PN-EN 13187:2001. Polish Standard: Infrared detection of thermal parameters of buildings.

Rozporządzenie Ministra Spraw Wewnętrznych i Administracji z dnia 04 lutego 1999 r. w sprawie warunków technicznych, jakim powinny odpowiadać budynki i ich usytuowanie. Dz. U. 1999 nr 15 poz.140 (Ordinance of Ministry of Internal Affairs and Administration, on Feb., 4, 1999: Technical parameters for new buildings – in Polish).

Rymsza B. 2003. *Biodeterioracja pleśniowa obiektów budowlanych (Mould biodererioration of buildings, in Polish)*. Published by Poznan University of Technology, Poznan.

Samson R.A., Hoekstra E.S., Frisvad J.C., 2004. *Introduction to food and airborne fungi*. Seventh Ed. Utrecht. Centralbureau voor Schimmercultures (CBS) The Netherlands.

Zyska B. 1996. Microbial disasters – a worldwide problem for the reliability data banks. In: Biodeterioration and Biodegradation Papers of the 10th International Biodeterioration and Biodegradation Symposium. Hamburg, DECHEMA Monographs 133, Frankfurt am Main, 427–432.

Management of Indoor Air Quality – Dudzińska (ed)
 ISBN 978-0-415-67266-5

A method for testing different materials for their use as radon barriers

K. Kozak
Institute of Nuclear Physics PAN, Kraków, Poland

ABSTRACT: The naturally occurring radioactive gas radon (Rn-222) enters dwellings mainly from soil. Therefore, currently the most important way to prevent contamination of indoor air with radon is via the application of special foils, which are placed between the soil and house foundations. These foils work as radon barriers.

The method for testing membranes for their usefulness as radon barriers in the building industry, was constructed at the Laboratory of Radiometric Expertise, Institute of Nuclear Physics PAN in Kraków. The special device (constructed in that Laboratory) consists of two chambers; the material to be tested is placed between them. Radon gas is pumped to one of the chambers and the concentrations are registered in both chambers using an AlphaGUARD ionization chamber. In this way the radon transmittance P and resistance Z through the foil are calculated and, based on these values and the thickness of the foil, the radon permeability k is determined. The results of test measurements of radon permeability for some types of foils used in the building industry are presented.

Keywords: ^{222}Rn, radon diffusion, radon permeability, membrane, radon mitigation

1 INTRODUCTION

Radon, a decay product of radium and a naturally occurring radioactive gas, is responsible for >40% of the annual dose of radioactivity received by the population in Poland (NAEA 2010). According to WHO (WHO 2009) radon is considered to be one of the single most important carcinogenic factors to which the population is exposed. It is important especially in connection with the indoor atmosphere in which people spend about 80% of their time. From this point of view, radon can be said to be the single component of indoor air which significantly influences its quality.

The main source of radon in dwellings is the soil and bedrock. Radon entry to a house and its indoor concentration are affected by many factors, e.g. air tightness of the substructure of the house, soil permeability, the infiltration characteristics of the house (Wang 2002).

Since radon enters the house mainly from the underlying soil, the insulation of foundations is crucial to protecting indoor air from radon. Special care should be taken when constructing houses in areas where geological faults occur. Swakoń et al. (Swakoń 2005) showed that higher radon concentration in soil gas was observed around local disjunctive tectonic zones. The simplest and relatively cheapest method to inhibit diffusion is to apply different types of membranes as radon gas barriers in building construction. Such membranes impede diffusion of the gas. Various insulating materials have been investigated as regards their ability to limit radon entry into houses. The studied materials include: vapour barrier membranes commonly used in building construction (Quindos Poncela 2005, Chen 2009), commercially available materials, e.g. polyethylene, aluminised polymers, makrofol, Plexiglas, paper, glass (Durcik 1996, Daoud 2001, Arafa 2002) and PVC, PE foils, bitumen membranes (Jiranek 2001).

A contribution to indoor radon can also come from building materials. Thus, some common plaster coating materials were also studied as a barrier to radon exhalation from walls (Singh 2005).

The facility constructed at the Laboratory of Radiometric Expertise (Institute of Nuclear Physics PAN, Krakow, Poland) and the method of measuring permeability of some commercial membranes is presented in the paper. This modified method is based on that drawn up at SP Swedish National Testing and Research Institute (SP Report 2006). The results of preliminary tests are discussed.

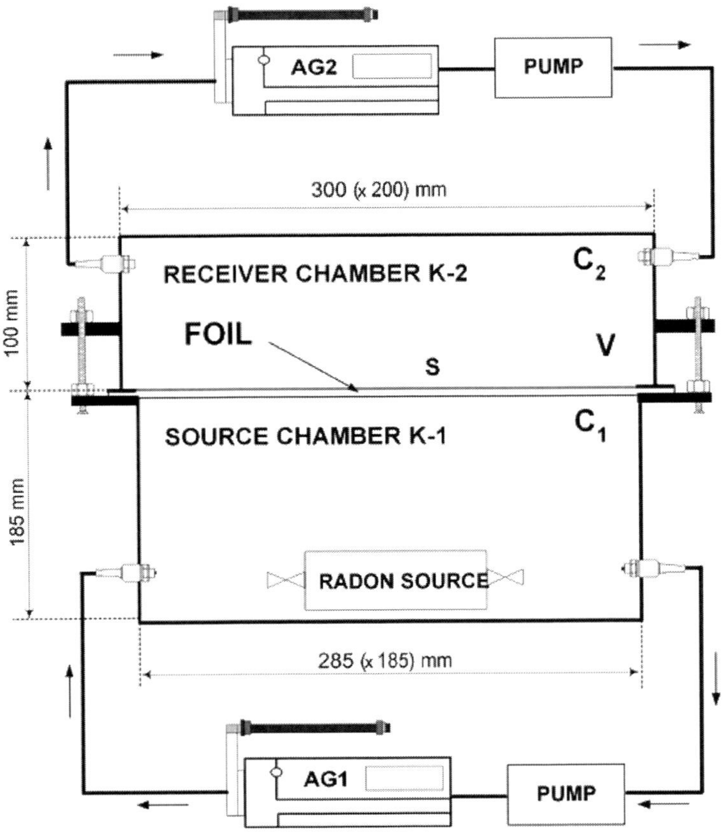

Figure 1. Schematic diagram of experimental setup.

2 MATERIALS AND METHODS

Figure 1 shows a scheme of the experimental setup. The device (called "RnBOX") consisted of two chambers K-1 (source: 9.8 dm^3) and K-2 (receiver: 6.0 dm^3). The tested material was placed between the chambers. The chambers were then flanged together, so that the connection between them was airtight. Radon gas was emitted from the source Rn-1025 (PYLON) into the K-1 chamber and a high concentration was allowed to generate there. Radon diffused through the membrane to be tested and the build-up of its concentration could be observed in the receiver chamber K-2. Two AlphaGUARD monitors (working in "flow mode" with flow rate of 0.3 dm^3/min) were used for continuous registration of radon concentrations in both chambers. The measurement lasted for about three days and was conducted in normal room conditions.

The radon transmittance was determined by measuring radon concentrations on both sides of the membrane. It was assumed that the radon concentration difference in both chambers is constant and that the effects of radon gas decay are negligible.

The radon flux density q [Bq/m^2s] is expressed by:

$$q = P \cdot (C_1 - C_2) \tag{1}$$

where: P – radon transmittance [m/s], C_1, C_2 – radon concentrations in K-1 and K-2 chambers, respectively [Bq/m^3].

The differential equation for radon concentration C_2 is:

$$\frac{dC_2}{dt} = \frac{P \cdot (C_1 - C_2) \cdot S}{V} \tag{2}$$

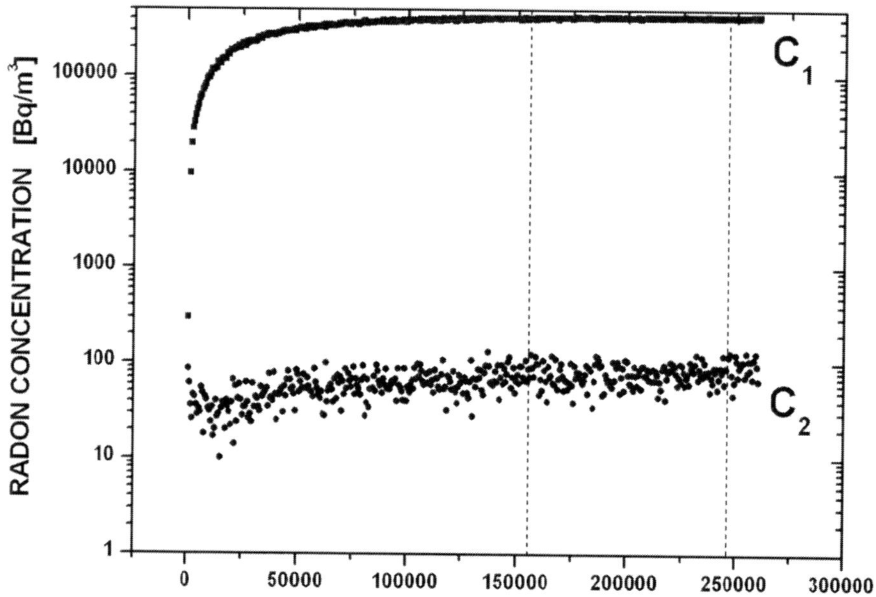

Figure 2. Difference in radon concentrations on both sides of tested material.

where: t – time [s], S – area of tested membrane [m^2], V – volume of receiver chamber K-2 [m^3].
Assuming the boundary conditions $C_2 = 0$ at $t = 0$ the solution of the Equation 2 can be written as:

$$C_2 = t \cdot \frac{S}{V} \cdot P \cdot (C_1 - C_2) \tag{3}$$

From (2) radon transmittance P can be determined:

$$P = \frac{C_2}{C_1 - C_2} \cdot \frac{V}{S \cdot t} \tag{4}$$

For a membrane made of homogenous material, the radon permeability k can be calculated according to the formula:

$$k = P \cdot d = \frac{C_2}{C_1 - C_2} \cdot \frac{V}{S \cdot t} \cdot d \tag{5}$$

where: k – radon permeability [m^2/s], d – membrane thickness [m].

Unlike the method applied at the SP Swedish National Testing and Research Institute, in our case the radon concentrations in chambers K-1 and K-2 were monitored continuously. This made it possible to observe the increase of radon concentration in both chambers and to choose the appropriate time interval when the radon concentration difference between both chambers was constant. The changes of radon concentration inside both chambers C_1 and C_2 are shown in Figure 2. The maximum concentration of ca. 420 kBq/m^3 was reached in source chamber K-1, whereas the concentration in the receiver chamber K-2 was almost stable at the level of 100 Bq/m^3.

For calculations, six pairs of values of C_1 and C_2 were chosen from the range of constant difference of radon concentration (marked with dashed vertical lines in Figure 2).

Then the radon transmittance P and radon permeability k were determined according to equations (4) and (5). An example of such calculations is shown in Table 1. Six values of radon permeability k were calculated taking into account pairs of radon concentrations (C_1, C_2) and the respective time. The final result of permeability k was the average of these six values.

Table 1. Example of calculations of radon transmittance and radon permeability through a membrane.

Radon Concentration $C1$	Radon Concentration $C2$	Time t	Radon transmittance P	Radon permeability k
	[Bq/m^3]	[s]	[m/s]	[m^2/s]
409600	99	159000	1.9 · 10-10	7.2 · 10-13
415744	67	180600	1.1 · 10-10	4.2 · 10-13
423936	91	198600	1.3 · 10-10	5.1 · 10-13
428032	72	223800	9.3 · 10-11	3.5 · 10-13
415744	116	234600	1.5 · 10-10	5.6 · 10-13
428032	128	259200	1.4 · 10-10	5.4 · 10-13

Table 2. Experimental values of radon permeability for five materials.

Material	Thickness d	Radon transmittance P	Radon permeability k
	[mm]	[m/s]	[m^2/s]
A	3.8	0.14 · 10-9	0.52 · 10-12
B	2.8	0.23 · 10-9	0.65 · 10-12
C	1.6	0.14 · 10-9	0.22 · 10-12
D	1.0	0.96 · 10-9	0.96 · 10-12
E	1.0	0.22 · 10-9	0.22 · 10-12

3 RESULTS AND DISCUSSION

Two commercial bitumen membranes were investigated as regards radon permeability using the set-up described. Additionally, three foils were studied within the intercomparison experiment organized by National Radiation Protection Institute SURO (Czech Republic). The results are presented in Table 2.

Errors of: radon concentration $C1$, $C2$, volume V and area S were estimated at $\pm 5\%$, thus the total uncertainty of measurement amounted to $\pm 15\%$.

The obtained values of permeability k range from $0.22 \cdot 10^{-12}$ to $0.96 \cdot 10^{-12}$ and are in agreement with the results reported by other authors. The Faculty of Civil Engineering, Czech Technical University tested radon permeability for more than 120 radon-proof materials obtained from around Europe. The results of radon diffusion coefficient varied from 10^{-15} to 10^{-8} m/s^2 (Jiranek 2004).

The materials A and B showed very good barrier properties against radon penetration. Material B is a bitumen membrane modified with an aluminium sheet.

Neither the manufacturers of the materials C, D and E nor their composition are known because of intercomparison experiment requirements. They were thinner than A and B, which resulted in higher radon permeability.

4 CONCLUSION

The paper presents a simple and relatively fast method for the assessment of radon diffusivity through a membrane. Thus, it can be used for testing the usefulness of commercially available foils as radon barriers. Preliminary test measurements for five materials have been carried out. The method seems to be sufficient for a first stage of investigation of different materials as radon retardants and for screening the best ones for more detailed examination.

Further study and improvement of the method is planned to better assess the barrier properties of the materials. The first property to be investigated will be the density on which radon diffusion is known to strongly depend. The radon diffusion coefficients for polyethylene membranes of the same thickness and different densities differ at about one order of magnitude (Jiranek 2004).

In a real building there is under-pressure, which causes the increase of radon entry through foundations (stack effect) and through radon-proof membranes. Experiments taking this into account are in planning.

Another important point is the membrane humidity because in real conditions the membranes work in a humid environment. It is planned to determine radon permeability for a membrane under different humidity conditions.

The membranes work as radon barriers together with the building foundation. Radon diffusion through combined samples (concrete and membrane) should be examined.

ACKNOWLEDGEMENTS

The author thanks his colleagues: Jadwiga Mazur and Dominik Grzadziel for their help in performing measurements and Tadeusz Zdziarski for technical assistance.

REFERENCES

Arafa, W. 2002. Permeability of radon-222 through some materials. *Radiation Measurements* 35: 207–211.

Chen, J., Ly, J., Schroth, E., Hnatiuk, S., Frenette, E. & Blain, M-F. 2009. Radon diffusion coefficients of vapour barrier membranes used in Canadian building construction. *Radiation and Environmental Biophysics* 48: 153–158.

Daoud, W.Z. & Renken, K.J. 2001. Laboratory assessment of flexible thin-film membranes as a passive barrier to radon gas diffusion. *The Science of the Total Environment* 272: 127–135.

Durcik, M. & Havlik, F. 1996. Experimental study of radon and thoron diffusion through barriers. *Journal of Radioanalytical and Nuclear Chemistry* 209(2): 307–313.

Jiranek, M. & Hulka, J. 2001. Applicability of various insulating materials for radon barriers. *The Science of the Total Environment* 272: 79–84.

Jiranek, M. 2004. Testing and design of radon resisting membranes based on the experience from the Czech Republic. *Proceedings of the 4th European Conference on Protection Against Radon at Home and at Work.*

NAEA (National Atomic Energy Agency) Annual Report 2009. Warszawa, 2010 (in Polish).

Quindos Poncela, L.S., Fernandez, P.L., Gomez Arozamena, J. & Sainz Fernandez, C. 2005. A method for measuring effective radon diffusion coefficients in radon barriers by using modified Lucas cells. *Radiation Measurements* 39: 87–89.

Singh, S., Singh, J., Singh, L. 2005. The study of some common plaster coating materials and plastic foils as a barrier to radon. *Radiation Measurements* 40: 673–677.

SP Report. 2006. *(private communication).*

Swakoń, J., Kozak, K., Paszkowski, M., Gradziński, R., Łoskiewicz, J., Mazur, J., Janik, M., Bogacz, J., Horwacik, T. & Olko, P. 2005. Radon Concentration in Soil Gas around Local Disjunctive Tectonic Zones in the Kraków Area. *Journal of Environmental Radioactivity* 78: 137–149.

Wang, F. & Ward, I.C. 2002. Radon entry, migration and reduction in houses with cellars. *Building and Environment* 37: 1153–1165.

WHO Handbook on Indoor Radon. A public health perspective. 2009. ed. Word Health Organisation, France.

Management of Indoor Air Quality – Dudzińska (ed)
© 2011 Taylor & Francis Group, London, ISBN 978-0-415-67266-5

Magnetic properties of dust as indicators of indoor air pollution: Preliminary results

M. Jeleńska, B. Górka-Kostrubiec & E. Król
Institute of Geophysics, Polish Academy of Sciences, Warsaw, Poland

ABSTRACT: This paper reports preliminary results of a study of the pollution of indoor air by Particulate Matter (PM) measured inside flats located in various places in Warsaw and in unpolluted suburbs. Air pollution was evaluated by measuring the magnetic susceptibility χ of dust taken from vacuum cleaners from polluted sites. Hysteresis parameters (H_{cr} – coercivity of remanence, H_c – coercivity, M_s – saturation magnetization and M_{rs} – saturation remanence), frequency dependent susceptibility χ_{fd} and thermal decay of saturation remanence (SIRM(T) curve) allow the identification of magnetic minerals and their granulometry. χ shows that the most polluted places are the city centre and the Ursynów district. Hysteresis parameters, χ_{fd} values and SIRM(T) curves for dust and for soil samples taken in the vicinity demonstrate that the indoor pollution contains finer particulates than outdoor pollution.

Keywords: air pollution, indoor dust, magnetic method

1 INTRODUCTION

Air pollutants present in urban dust are released from fuel combustion in vehicle traffic through metal wear in abrasion/corrosion processes, from domestic sources (heating fires) and as products of industrial activity. Air pollutants contain iron-rich magnetic particles which consist of different phases of iron oxides such as magnetite, maghemite, hematite or mixtures of these phases. These magnetic particles penetrate to flats from outside and contribute to indoor dust.

Several studies applied magnetic techniques to identify atmospheric pollution by Particulate Matter (PM) collected on filters (e.g. Maxworthy et al. 2001, 2002, 2003, Spassov et al. 2004; Sagnotti et al. 2006, 2009), deposited on tree leaves (e.g. Michell et al. 2009, Maher et al. 2008) and street dust (Shilton et al. 2005, Bućko et al. 2009). Identification of magnetic mineral composition and structure can be used to discriminate the source of pollution (Spassov et al. 2004) and help to monitor the distribution of anthropogenic and natural PM in time and space.

We carried out a magnetic study of pollution of soils from Poland, Ukraine and Slovakia (Jeleńska et al. 2004) and pollution of air collected on filters in relation to weather conditions in Warsaw in the 1980s, 1990s and 2010s (Król et al. 2009). The study provides information about the current and former state of magnetic PM pollution in Warsaw. We found that the annual trend of the magnetic susceptibility changes depends on local meteorological conditions such as temperature, humidity and precipitation. In parallel, we have examined the pollution of soil along highly frequented-traffic roads in Warsaw using the magnetic method. We found that the volume of traffic is one of the factors responsible for the transport of high magnetic susceptibility (concentration of anthropogenic magnetic particles) into the roadside soil at a distance of 3 m from the wayside.

Pollution of the indoor environment in relation to outdoor and indoor sources was studied by Kingham et al. (2000), by Fischer et al. (2000) and by He et al. (2004). They examined the PM_{10} and $PM_{2.5}$ together with chemical pollutants such as Polycyclic Aromatic Hydrocarbons (PAH) and Volatile Organic Compounds (VOC). Kingham et al. (2000) found that neither of these parameters are reliable indicators for traffic-related air pollutants. However, they found a significant correlation between outdoor and indoor concentrations of pollutants. The results were confirmed by Fischer et al. (2000). Halsall et al. (2008) used a magnetic method combined with PAH to study the relation between indoor and outdoor pollution. Magnetic properties of cigarette ashes were examined by Jordanova et al. (2006). They reported that cigarette ashes contain a large amount of a mixture of very fine and coarse magnetite grains.

Magnetic and aerobiological studies of indoor and outdoor dust from Bulgaria were presented in Jordanova et al. (2010). Their investigations of magnetic properties of indoor and outdoor dust suggest the presence of an anthropogenic magnetic fraction containing magnetite, maghemite and hematite. The study of heavy metal concentration in house dust from Warsaw was reported by Tatur et al. (2006), however the authors did not examine the spatial distribution of contamination.

In this paper we present the application of the magnetic method to investigate magnetic properties of indoor air pollution. The magnetic method is particularly sensitive to iron-rich pollutants which mostly arise from vehicle traffic and are related to outdoor sources. As a first step, the aim of our study is limited to magnetic properties of indoor dust.

2 DESCRIPTION OF EXPERIMENTAL METHODS AND SAMPLES

The magnetic method is based on measurements of the magnetic susceptibility (χ), Anhysteretic Remanent Magnetization (ARM), hysteresis parameters (saturation magnetization M_s, saturation remanence M_{rs}, coercivity H_c, coercivity of remanence H_{cr}) and frequency dependent χ defined as $\chi_{fd} = (\chi_1 - \chi_2)/\chi_1$ where χ_1 and χ_2 are the magnetic susceptibility measured in low and high frequency field, respectively.

Magnetic susceptibility is used as a detector of magnetic particle concentrations in pollution and could serve as the initial monitoring of pollution to indicate places for detailed analysis by traditional chemical method and by urban monitoring of PM concentration. The frequency dependence of susceptibility determines the granulometry of magnetic minerals. Identification of magnetic minerals and their granulometry were made by measuring the decay curve of SIRM (isothermal saturation remanence) during heating and by determination of magnetic hysteresis parameters. Magnetic techniques are very sensitive to the presence of magnetic particles in pollution, non-destructive to samples and very simple and cheap for applications.

The dust samples were taken from the vacuum cleaner bags of flats located in twelve places in Warsaw distributed in such polluted places as the city centre or locations near by major roads with heavy traffic, and in unpolluted suburbs. The locations were differentiated by the floor where flats are situated (Tab. 1). Dust samples were sieved and placed in plastic cubic boxes and weighed before measurement.

The magnetic susceptibility of dust samples was measured in the paleomagnetic laboratory of the Institute of Geophysics in Warsaw using the KLY-2 kappa-bridge made by the Geofizyka–Brno (Czech Republic). Magnetic susceptibility was normalized by mass (χ in the unit $10^{-8}\,m^3\,kg^{-1}$). The value was the mean computed from three repeated measurements.

The frequency dependence of magnetic susceptibility was determined by measuring the susceptibility at two frequencies: 976 Hz and 3900 Hz of magnetic field using the AGICO MFK1-FA Multi-Function-Kappa Bridge.

Hysteresis properties (H_c, H_{cr}, M_s, M_{rs}) were measured in a maximum field of 1T using the VSM magnetometer produced by the Molspin (Great Britain).

Continuous thermal demagnetization of saturation isothermal remanence (SIRM) was carried out with the use of a device made by the TUSK, Poland. A sample was magnetized in a magnetic field of 9T, high enough to saturate hard magnetic minerals such as hematite or goethite. Remanence was measured during heating to 700°C in a magnetic screen. Identification of magnetic minerals was made by determination of unblocking temperature (T_{ub}).

3 RESULTS AND DISCUSSION

Results of magnetic properties for samples of the indoor dust measured at the 12 flats in Warsaw are listed in Table 1 and values of magnetic susceptibility are shown in Figure 1.

The values of susceptibility ranged from 22 to $206 \times 10^{-8}\,m^3/kg$ and are comparable with the data ($57 - 1700 \times 10^{-8}\,m^3/kg$) obtained by Jordanova et al. (2010) for Burgas and Sofia in Bulgaria except extremely high values which we do not observe.

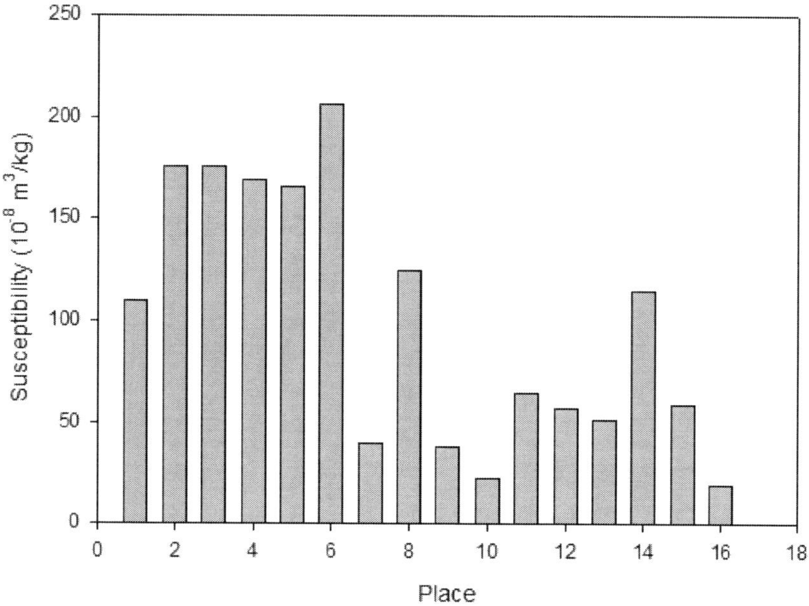

Figure 1. Distribution of magnetic susceptibility for samples of indoor dust measured at flats in Warsaw. Number of sampling sites as in Table 1.

The highest values of susceptibility were measured in the flats located in the city center in Figure 1 marked as C-3, C-4, C-5 and C-6. The same high values are observed in Ursynów (place U-2) from the flat facing busy street. The second flat from Ursynów (place U-1), located far-away from the busy streets, shows a smaller value of χ. One high value of χ is also observed for the flat from Sluzew (S-8). The flat is placed on the ground floor and faces a very narrow and crowded street. The last high value of χ was observed for Bialoleka (B-14), a suburb of Warsaw with relatively fresh air. The flat is located on the ground floor and separated from the local street by a small garden. For this location, a natural high anomaly of the Fe content in soil was observed (Areas of air pollution by heavy metals in Warsaw – 2003 r.) so we believe that in this case the high value of χ is related to a natural source. The reliability of the data was checked by measuring two dust samples taken from the same flat (places C-4 and C-5) and for the dust taken from two flats from the same building on the same floor, one facing a street and the second one facing a yard. The differences in values of χ for these flats are not significant.

The two last values of χ shown in Figure1 are for soil samples taken at the sides of two busy main streets. Although the values are comparable with those for less polluted flats, direct comparison of χ for dust and soils is not possible as the mode of pollution deposition is different in both cases. However, we can compare magnetic mineralogy of dust samples and soil samples.

SIRM(T) curves (Fig. 2) indicate that the magnetite of the unblocking temperature $T_{ub} = 575°C$ is the main magnetic mineral in samples of indoor dust.

Besides magnetite, two inflexion points are visible on the heating curves, one around 150°C and a second one around 380°C. We suppose that the inflexion point at around 150°C is evidence of the presence of fine meta-stable single domain (SD) grains of magnetite close to the superparamagnetic (SP) state. The second inflexion point around 380°C can be connected with maghemite which converts at that temperature to hematite. Sometimes, also a small hematite tail was observed. For soil samples the first inflexion point is not seen and the second point is broader and less visible, and hematite is well pronounced. This suggests that in soil, besides hematite, coarser magnetite/maghemite grains are present. A curve similar to those for soil is observed for one dust sample (Ursynów, U-1). The same mineralogy is reported by Jordanova et al. (2010) for dust from Bulgaria. Similarly results were also reported by Sagnotti et al. (2006) in a study of magnetic particles in the PM10 air pollution in the Latium Region (Italy). The low-coercivity, magnetite-like minerals were identified in urban air pollution, and this result is consistent with former reports

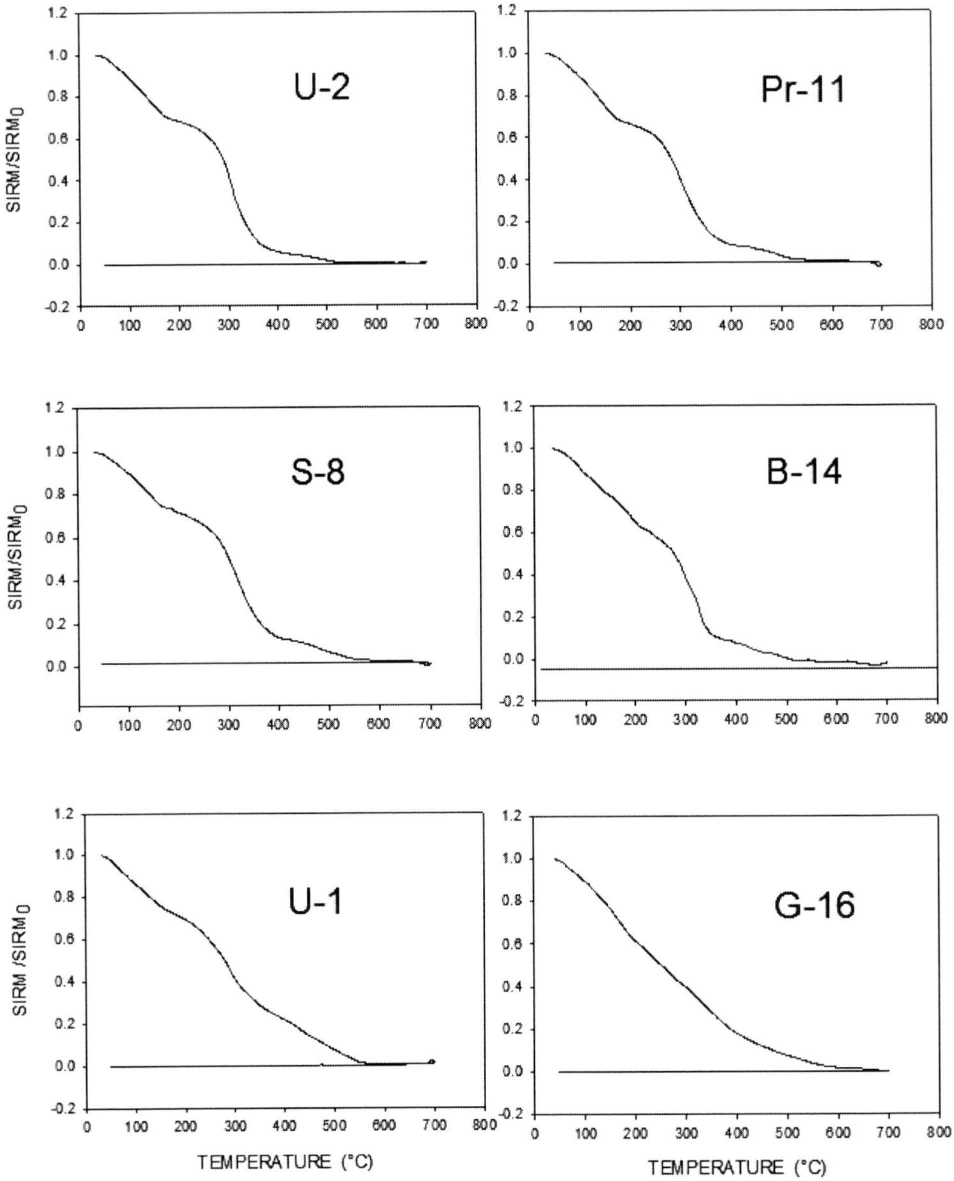

Figure 2. Examples of decay curves of SIRM(T) for dust samples from different districts of Warsaw. Number of sampling sites as in Table 1.

from other European metropolises: Munich (Maxworthy et al. 2001, 2002, 2003), Zurich (Spassov et al. 2004) and Madrid (Gomez-Paccard et al. 2004).

The frequency-dependence of magnetic susceptibility χ_{fd} is used as an indicator of the presence of fine ferromagnetic grains in a sample. For a grain size close to SP/SD boundary χ_{fd} is >8, usually between 10 and 12, for multidomain (MD) grains χ_{fd} is less than 6, and for very fine grains χ_{fd} is again very small (Dearing et al. 1997, Maher & Taylor 1988). χ_{fd} values are less than 2 for dust samples (Tab. 1).

The ratios of hysteresis parameters M_{rs}/M_s and H_{cr}/H_c display on the Day plot (Day 1977) also give some information about grain size and domain state of magnetic particles. Dunlop (Dunlop 2002a, b) modified the Day plot by considering the influence of SP grains. As can be seen in

Table 1. Description of localities of dust sampling sites and their magnetic parameters: susceptibility (χ), frequency dependency of susceptibility (χ_{fd}) and hysteresis parameters (H_c, H_{cr}, M_s, M_{rs}).

Sample nr.	Locality	χ (10^{-8} m^3/kg)	H_c (mT)	H_{cr} (mT)	M_{rs}/M_s	H_{cr}/H_c	χ_{fd} (%)
1 (U-1)	Ursynów, Grzegorzewskiej st. (2nd floor)	110.1	7.0	52.0	0.094	7.5	4.17
2 (U-2)	Ursynów, Herbsta st. (3rd floor)	176.0	5.0	22.0	0.056	4.4	0.73
3 (C-3)	Centre, Andersa st. (3rd floor)	175.6					0.54
4 (C-4)	Centre, Grzybowska st. (15th floor) a) deep	169.15	8.5	27.0	0.075	3.2	1.36
5 (C-5)	Centre, Grzybowska st. (15th floor) b) shallow	166.1					0.64
6 (C-6)	Centre, Wyzwolenia av, (3rd floor)	206.2	6.0	33.0	0.06	5.5	1.85
7 (M-7)	Mokotów Bruna st. (5th floor)	40.1					1.12
8 (S-8)	Służewiec Etiudy Rewolucyjnej st. (ground floor)	124.8	6.5	44.0	0.068	6.8	0.80
9 (Po-9)	Powiśle Czerniakowska st. (3rd floor)	37.8					1.66
10 (Wi-10)	Wilanów Sobieskiego st. (3rd floor)	22.1					?
11 (Pr-11)	Praga south Umińskiego st. (1st floor) street	64.3	7.5	26.0	0.088	3.5	1.38
12 (Pr-12)	Praga south Umińskiego st. (1st floor) yard	57.0					1.27
13 (Wo-13)	Wola Pustoła st. (4th floor)	51.6					0.99
14 (B-14)	Białołęka Ostródzka st. (ground floor)	114.9	7.0	24.5	0.073	3.5	1.33
15 (G-15)	Soils Żerań Wsch.	58.5	10.5	30.5	0.15	2.9	1.29
16 (G-16)	Soils Ursynów Rosoła st.	19.2	8.5	23	0.16	2.7	1.48

Figure 3 soil samples lie on the mixing curve of SD + MD grains. Dust samples lie on or in the proximity of the same mixing curve with the exception of two samples which are shifted towards higher values of H_{cr}/H_c ratio.

Interpreting the results of magnetic experiments is not obvious. SIRM(T) curves point to presence of magnetite/maghemite with small contribution of hematite. Inflexion points at temperature around 150°C indicate a contribution of fine unstable SD grains of magnetite-like minerals. χ_{fd} less than 2 (except U-1) can be related to very fine SP grains (Dearing et al. 1997, Maher & Taylor 1988). Hysteresis parameters (Fig. 3) point to a contribution of SD and MD grains. In fact, each experiment reveals the importance of grains in different domain states. At the beginning we examined the influence of SP grains suspecting their presence from χ_{fd} values. They do not contribute to SIRM(T) curves. The presence of SP grains enhances susceptibility and diminishes H_c value, whereas H_{cr} remains unchanged. As a result H_{cr}/H_c decreases. The relationships of H_c and H_{cr}/H_c with χ (Fig. 4) show that H_c increases and H_{cr}/H_c decreases with increase of χ values – this point to an increasing contribution from SP grains. Three samples: U-1, S-8 and C-6 do not show the relationship H_{cr}/H_c vs. χ observed for other samples. These samples exhibit high values of H_{cr} indicating that rather different magnetic mi neralogy is more important than the SP contribution. However, SIRM(T) curves show the same magnetic mineralogy for all samples except U-1. We are not able to explain this paradox. If we omit these samples we can propose that $\chi_{fd} < 2$ obtain for indoor dust samples characterizes SP grains. The studies (Spassov et al. 2004, Sagnotti et al. 2009) of magnetic properties

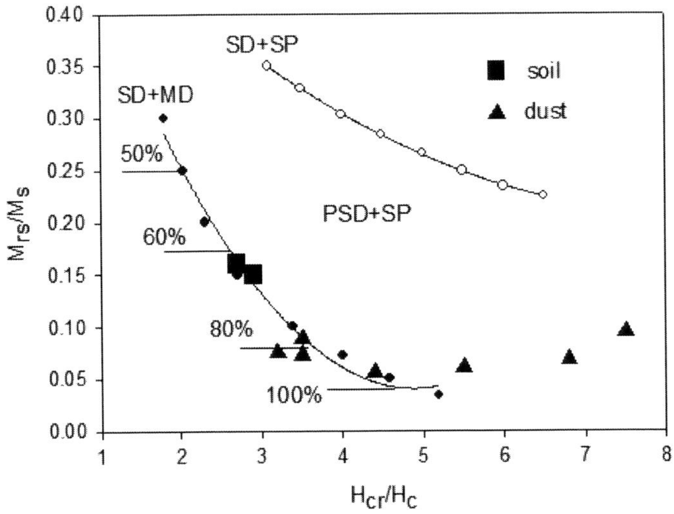

Figure 3. Hysteresis ratios M_{rs}/M_S and H_{cr}/H_C for dust samples shown in the Day diagram (Day 1977) with mixing curves calculated theoretically by Dunlop (Dunlop 2002a, 2002b). Full circles – SD+MD mixing curve; open circles – SD+SP (10 nm) mixing curve. Full triangles – dust samples; full squares – soil samples.

of PM accumulated on leaves and air filters in Rome show that the values of χ_{fd} in the range 9–11% characterize magnetic grains from brakes, the values in the range 7–8% characterize magnetic grains from diesel exhaust pipes and the values in the range 1.5–2% characterizes magnetic grains from gasoline exhaust pipes. We can suppose that dust collected in Warsaw flats can originate from gasoline exhaust pipes, however, the "gasoline exhaust" particles show higher coercivity than our dust samples.

Sagnotti et al. (2009) showed that magnetic minerals collected on air filters are combination of different grain populations derived from abrasion of disk brakes, and diesel and gasoline exhaust pipes produced by vehicle traffic. The SP fraction contributes to a mixture of SD, PSD and MD particles in variable proportion and is mostly produced by disk brakes. We suppose that what we observe in indoor dust is a mixture of SD and MD grains with a contribution from SP grains.

4 CONCLUSIONS

The aim of this study is to examine the influence of outdoor pollution derived from vehicular traffic on indoor air quality by means of a magnetic method. We did not consider indoor sources of pollution such as cooking, fire places, smoking or presence of pets and children. We did not compare magnetic pollution with chemical or biological pollution. We focused our interest on iron-rich particles related to vehicle traffic. The preliminary results of a study of dust taken from various flats in Warsaw show that magnetic susceptibility is a good proxy for monitoring of indoor air pollution. The method should be developed and improved towards more unified sampling to obtain more comparable results. Some conclusions can be drawn even from such preliminary data:

1. measurements of magnetic susceptibility of dust samples show that the flats located in the center of Warsaw and in the Ursynów district are the most polluted ones,
2. the level of the floor of a flat with respect to street level has no influence on the pollution extent or composition,
3. the ground floor in a narrow and crowded street is the most dangerous location,
4. magnetic minerals present in indoor dust are a mixture of SD and MD grains with contribution of SP grains.

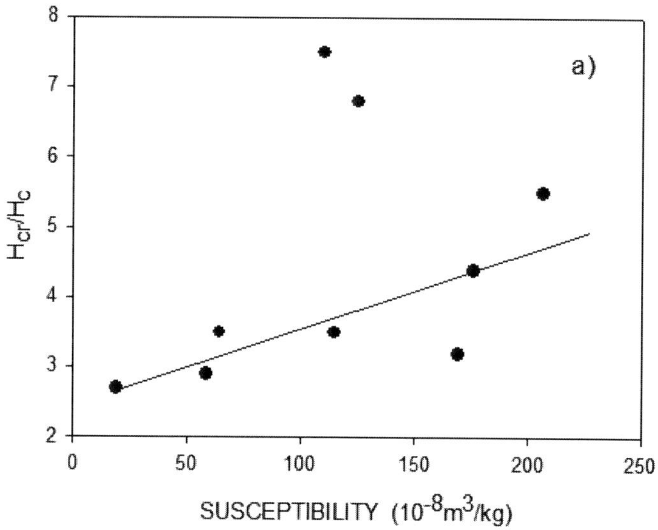

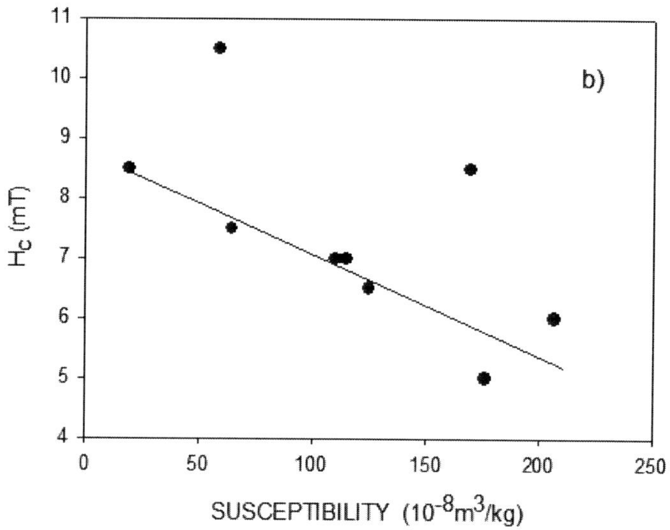

Figure 4. Susceptibility versus H_{cr}/H_c ratio (a) and H_c (b).

REFERENCES

Biuro Naczelnego Architekta Miasta, Opracowanie Ekofizjograficzne do Studium Uwarunkowań i Kierunków Zagospodarowania Przestrzennego m. st. Warszawy – 2003. Rejony Zanieczyszczenia Powietrza Metalami Ciężkimi w Warszawie – 2003.

Bućko, M.S., Magiera, T., Pesonen, L. J., & Janus, B. 2009. Magnetic, Geochemical, and Microstructural Characteristics of Road Dust on Roadsides with Different Traffic Volumes – Case Study from Finland. *Water Air Soil Pollut.* DOI 10.1007/s11270-009-0198-2.

Day, R., Fuller, M. D. & Schmidt, V. A. 1977. Hysteresis properties of titanomagnetites: Grain size and composition dependence. *Phys. Earth Planet. Inter.* 13: 260–266.

Dearing, J.A., Bird, P.M., Dann, R.J.L. & Benjamin, S.F. 1997. Secondary ferrimagnetic minerals in welsh soils: a comparison of mineral magnetic detection methods and implication for mineral formation. *Geophys. J. Int.* 130: 727–736.

Dunlop, D. J. 2002a. Theory and application of the Day plot (Mrs/Ms versus Hcr/Hc): 1. Theoretical curves and tests using titanomagnetite data. *J. Geophys. Res.* 107(B3): 2056, DOI: 10.1029/2001JB000486.

Dunlop, D. J. 2002b. Theory and application of the Day plot (Mrs/Ms versus Hcr/Hc): 2. Application to data for rocks, sediments, and soils. *J. Geophys. Res.* 107(B3): 2057, DOI:10.1029/2001JB000487.

Fischer, P.H., Hoek, G., Van Reeuwijk, H., Briggs, D.J., Lebret, E., Van Wijnen, J.H., Kingham, S. & Elliott P.E. 2000. Traffic-related differences in outdoor and indoor concentrations of particles and volatile organic compounds in Amsterdam. *Atmospheric Environment* 34: 3713–3722.

Gomez-Paccard, M., McIntosh, G., Villasante, V., Osete, M. L., Rodriguez-Fernandez, J. & Gomez-Sal J. C. 2004. Low-temperature and high magnetic field measurements of atmospheric particulate matter. *J. Magn. Magn. Mater.* 272: 2420–2421.

Halsall, C.J., Maher, B.A., Karloukovski, V.V., Shah, P. & Watkins S.J. 2008. A novel approach to investigating indoor/outdoor pollution links: Combined magnetic and PAH measurements. *Atmospheric Environment* 42: 8902–8909.

He, C., Morawska, L., Hitchins, J. & Gilbert, D. 2004. Contribution from indoor sources to particle number and mass concentrations in residential houses. *Atmospheric Environment* 38: 3405–3415.

Jeleńska, M., Hasso-Agopsowicz, A., Kopcewicz, B., Sukhorada, A., Tyamina, K., Kądziałko-Hofmokl, M. & Matviishina, Z. 2004. Magnetic properties of the profiles of polluted and non-polluted soils. A case study from Ukraine. *Geophys. J. Int.* 159: 104–116.

Jordanova, N., Jordanova, D., Yankova, R., Petrov, P., Popov, T. & Tsacheva, T. 2006. Magnetism of cigarette ashes. *J. Magn. Magn. Mater.* 301: 50–66.

Jordanova, N., Jordanova, D., Yankova, R., Petrov, P., Popov, T. & Tsacheva, T. 2010. Magnetic and aerobiological studies of indoor and outdoor dust from Bulgaria. List of abstracts. *12 Castle Meeting on Paleo, Rock and Environmental Magnetism*.strony?

Kingham, S., Briggs, D., Elliott, P., Fischer, P. & Lebret, E. 2000. Spatial variations in the concentrations of traffic-related pollutants in indoor and outdoor air in Huddersfield, England. *Atmospheric Environment* 34: 905–916.

Król, E., Jeleńska, M. & Górka-Kostrubiec, B. 2009. Magnetic parameters of membrane filters of the air – as the control factors of the iron compounds air pollution in Warsaw. *Int. Symposium on soil, sediment and dust magnetism, Bytom, Abstracts*: 30.

Maher, B.A., Mooreb, C. & Matzka, J. 2008. Spatial variation in vehicle-derived metal pollution identified by magnetic and elemental analysis of roadside tree leaves. *Atmospheric Environment* 42: 364–373.

Maher, B.A. & Taylor, R.M. 1988. Formation of ultrafine-grained magnetite in soil. *Nature* 336: 368–370.

Mitchell, R. & Maher, B. 2009. Evaluation and application of biomagnetic monitoring of traffic-derived particulate pollution. *Atmospheric Environment* 43: 2095–2103.

Muxworthy, A. R., Matzka J. & Petersen, N. 2001. Comparison of magnetic parameters of urban atmospheric particulate matter with pollution and meteorological data. *Atmospheric Environment* 35: 4379–4386.

Muxworthy, A. R., Matzka, J., Davila, A.F. & Petersen, N. 2003. Magnetic signature of daily sampled urban atmospheric particles. *Atmospheric Environment* 37: 4163–4169.

Muxworthy, A. R., Schmidbauer, E. & Petersen, N. 2002. Magnetic properties and Mössbauer spectra of urban atmospheric particulate matter: a case study from Munich, Germany. *Geophys. J. Int.* 150: 558–570.

Sagnotti, L., Macri, P., Egi, R. & Mondino, M. 2006. Magnetic properties of atmospheric particulate matter from automatic air sampler stations in Latinum (Italy): Toward a definition of magnetic fingerprints for natural and anthropogenic PM_{10} sources. *J. Geophys. Res.* 111: B12S22. DOI: 10.1029/2006JB004508.

Sagnotti, L., Taddeucci, J., Winkler, A. & Cavallo, A. 2009. Compositional, morphological, and hysteresis characterization of magnetic airborne particulate matter in Rome, Italy. *Geochem. Geophys. Geosyst.* 10: Q08Z06, DOI: 10.1029/2009GC002563.

Shilton, V. F., Booth, C. A., Smith, J. P., Giess, P., Mitchell, D. J. & Williams, C. D. 2005. Magnetic properties of urban street dust and their relationship with organic matter content in the West Midlands, UK. *Atmospheric Environment* 39: 3651–3659.

Spassov, S., Egli, R., Heller, F., Nourgaliev, D.K. & Hannam, J. 2004. Magnetic quantification of urban sources in atmospheric particulate matter. *Geophys. J. Int.* 159: 555–564.

Tatur, A., Gromadka, P. & Wasiłowska, A. 2006. Heavy Metals in House Dust from Warsaw. *Ecological Chemistry and Engineering.* 13(7): 695–702.

Management of Indoor Air Quality – Dudzińska (ed)
 ISBN 978-0-415-67266-5

TiO_2-based photocatalysts in indoor swimming pool air purification

A. Hänel, A. Korkosz & J. Hupka
Department of Chemical Technology, Chemical Faculty, Gdańsk University of Technology, Gdańsk, Poland

ABSTRACT: Modern swimming pools and aqua parks present special challenges to air treatment techniques. Firstly, airborne microorganisms pose several health threats. Secondly, Disinfection By-Products (DBPs) are present, which might adversely influence bathers and pool personnel. The hazards of microorganisms and DBP's are reviewed. Furthermore, the paper highlights the photodegradation of volatile chlorinated compounds and airborne microorganisms in water-based recreational facilities. Photoreactors and the operating conditions thereof are summarized from the point of view of scientific and practical applications.

Keywords: TiO_2, photoreactor, chlorinated VOC, chlorinated DBP, air born germs

1 INTRODUCTION

1.1 *Disinfection*

Chlorine is most commonly used for swimming pool water treatment to kill waterborne pathogenic organisms or to render them inert. The WHO guidelines recommend a pH range for swimming pools between 7.2 and 7.8, which has to be adjusted by the addition of alkali (e.g. sodium carbonate) or acidic solutions (e.g. sodium hydrogen sulfate) (WHO 2006). For Poland, it exists only a recommendation for the preferred pH, which is based on the German DIN 19643/97. The pH should be between 6.5 and 7.8 (DIN Norm 1997). The preferred chlorine concentration (defined as the sum of HOCl, OCl^- and $Cl_{2(aq)}$ concentration) ranges between 1–5 mg/dm^3 as Cl_2 (WHO 2006). Free chlorine level of 1 mg/dm^3 has been shown as adequate for routine disinfection. Lower concentrations ($\leq$0.5 mg/dm^3) are only recommended when chlorine is used in combination with ozone or UV disinfection (WHO 2006). However, the chlorine also reacts with sweat, urine and other organic compounds introduced by bathers and leads to the formation of halogenated DBPs.

Some halogenated DBPs migrate from the aqueous phase into the gas phase. Weaver et al. (2009) identified several volatile halogenated DBPs, later called chlorinated volatile organic compounds (VOC), which may have health effects on swimmers and employees. The identified chlorinated volatile DBPs are monochloroamine (NH_2Cl), dichloroamine ($NHCl_2$), trichloroamine (NCl_3), chloroform ($CHCl_3$), cyanogen chloride (CNCl), dichloroacetonitrile ($CNCHCl_2$), and dichloromethylamine (CH_3NCl_2) (Nemery et al. 2002, Weaver et al. 2009), whereby chloroform is present in the greatest concentration (WHO 2006). The chloramines are produced by substitution reactions and depend on the initial chlorine to nitrogen (Cl:N) molar ratio and pH (see Equations 1–3) (Li & Blatchley III 2009). The reaction equilibrium is reached rapidly for all equations.

$$NH_3 + HOCl \longleftrightarrow NH_2Cl + H_2O \tag{1}$$

$$NH_2Cl + HOCl \longleftrightarrow NHCl_2 + H_2O \tag{2}$$

$$NHCl_2 + HOCl \longleftrightarrow NCl_3 + H_2O \tag{3}$$

Monochloroamine is primarily formed in the pH range of 6.5 to 8.5 and at chlorine to ammonia ratio $\leq$1. The breakpoint, where almost all ammonia-nitrogen is oxidized to nitrogen, nitrite, nitrate and other products, occurs at the Cl:N ratio of approximately 1.6–1.7 (Li & Blatchley III 2009). Above the breakpoint, the predominantly formed chloroamine is NCl_3. NH_4^+ concentration in swimming pools is typically found in the order of 10 μg/dm^3, which is well beyond the Cl:N ratio of 1.7. Thus, trichloroamine is prevalently encountered.

1.2 *Threats from chlorination by-products*

Halogenated DBP gases can cause ocular and respiratory symptoms, asthmatic conditions as well as skin diseases for bathers and employees (Fantuzzi et al. 2010). Especially trichloroamine is the source of "chlorine" smell and tissue irritation (Hery et al. 1995) and it is believed that it acts as a lung sensitizer (Jacobs et al. 2007). Jacobs et al. measured the chloroamine concentration and evaluated the questionnaires of 624 pool workers in 38 Dutch swimming facilities (Jacobs et al. 2007). The average trichloroamine level was 0.66 mg/m^3 and exceeded the postulated comfort level of 0.50 mg/m^3. Poolside workers showed higher risk to develop upper respiratory symptoms and these symptoms were statistically significantly associated with cumulative chloroamine levels. Fantuzzi et al. (2010) investigated the prevalence of self-reported respiratory, ocular and cutaneous symptoms in subjects working at indoor swimming pools. 133 employees in twenty indoor swimming pools in the Emilia Romagna region of Italy were questioned and the trihalomethane concentration in alveolar air samples was evaluated. Subjects with mean values of trihalomethanes above 21 μg/m^3, like lifeguards and trainers (28.5 ± 20.2 μg/m^3), showed more frequently red eyes, runny nose, voice loss and cold symptom than other employees like subjects working in café areas, offices and engine rooms. In addition, verrucas, mycosis, eczema and rash were more experienced by lifeguards and trainers than for other workers. Recently, results of a cross-sectional survey of schoolchildren have shown that pulmonary damage and possible asthma are more often for children swimming in chlorinated pools. Their serum concentrations of pneumo proteins was higher than of control children which did not attend frequently swimming pools (Nemery et al. 2002).

1.3 *Microbiological hazards*

Preventing microorganisms in air is by far more complex than for water, where disinfectants are added and are present over a certain time. High relative humidity, characteristic for not well-maintained water-based recreational facilities, abets mold fungi growth. Aerosols from natural spas, hot tubes, whirlpools, HVAC systems and showers may be a source of airborne microorganisms.

Several potential pathogenic bacteria were detected in the air where *Legionella* spp. and *Mycobacterium* spp. are reported in the highest frequency. Besides *Legionellae* and *Mycobacteria* other pathogenic bacteria were detected, like *Staphylococcus hominis*, *Streptococcus pneumoniae*, *P. melaninogencia*, *A. otitis* and *Gemella morbillorum* (Angenent et al. 2005). Also viruses, fungal spores and microbial metabolism byproducts can be present in air and cause allergic responses, infectious diseases, respiratory problems and hypersensitivity reactions (Chen et al. 2010). Leoni et al. studied the hygienic conditions in 12 swimming pools in the city of Bologna and concluded that showers pose higher threat of *Legionella* spp. than the actual swimming pools (Leoni et al. 2001). *Legionella* spp. is Gram-negative bacteria, which live within amoebae and other protozoa or within biofilms. *Legionella* spp. can cause legionellosis whereat ninety percent of cases can be traced back to *L. pneumophila*. The disease shows symptoms of pneumonia or influenza. The Legionnaires' disease is one form with pneumonic symptoms. Pontiac fever is a non-pneumonic, non-fatal, influenza-like illness.

Mycobacterium spp. is rod-shaped bacteria, which may cause hypersensitivity pneumonis, pneumonia and skin infections. The cell wall is hydrophobic due to its high lipid content. Thus, the bacteria concentrate preferentially on gas bubble surfaces and migrate from aqueous to gas phase. *Mycobacteria* is also more resistant to disinfection due to the high lipid containing cell wall (WHO 2006).

2 FATE OF CHLORINATION BY-PRODUCTS

2.1 *Photochemical reactions of halogenated compounds in air*

Halogenated compounds decompose under the irradiation with UV light. The rate of reaction can be accelerated by photocatalyst. TiO_2 is the most widely used photocatalyst. The band gap of anatase

TiO_2 amounts 3.2 eV and if irradiated by light with $\lambda \leq 388$ nm, electrons are transferred from the valence into the conduction band forming electron-hole-pairs (Eq. 4).

$$TiO_2 + hv \rightarrow TiO_2(e^- + h^+) \quad (4)$$

The electron holes in the valence band can oxidize adsorbed water and/or hydroxyl groups which yields in the formation of reactive hydroxyl radicals (•OH). The electron in the conduction band reduces adsorbed oxygen under the formation of superoxide radicals ($O_2^{\bullet -}$) which can further react to hydroxyl radicals. Furthermore, electrons and electron-holes can directly react with adsorbed organic compounds. All mentioned radical reactions led to the decomposition of adsorbed organic compounds.

Complete mineralization of volatile chlorinated DBPs forms H_2O, CO_2 and HCl. The mechanism of DBP decomposition by TiO_2 is not fully understood yet, but three initializing reactions are proposed (Demeestere et al. 2007). Hydroxyl radicals (•OH) can initiate the decomposition (Eq. 5).

$$RH_xCl_y + \bullet OH \rightarrow \bullet RH_x - 1Cl_y + H_2O \quad (5)$$

Chlorinated alkanes like trichloromethane form chlorinated alkyl radicals by •OH induced hydrogen abstraction (Eq. 5). The formed chlorinated alkyl radicals react with molecular oxygen under formation of CO_2, HCl, and Cl_2. For unsaturated DBPs the preferred reaction of the •OH radical is the addition to the C=C bound (Hamill et al. 2001) (Eq. 6).

$$R_1C = CR_2 + \bullet OH \rightarrow R_1(OH)C - \bullet CR_2 \quad (6)$$

Hegedüs and Dombi (2004a, b) proposed for chlorinated ethenes that they get directly oxidized by electron-holes and $O_2^{\bullet -}$ radicals. Whereby Hwang et al. (Hwang et al. 1998) proposed that formed $O_2^{\bullet -}$ radicals attack the C=C double bound.

Yamazaki et al. observed deterioration of the photocatalyst decomposing trichloromethane (Yamazaki et al. 2005). They explained the deactivation by accumulation of Cl^- at the TiO_2 surface. Washing of the photocatalyst with water recovered the catalytic activity from its deterioration.

Feiyan et al. (2002) investigated the photodecomposition kinetics for chloroform, carbon tetrachloride and trichloroethylene without photocatalyst (low-pressure mercury lamp with 92% output at 254 nm and 6% at 185 nm). $Cl^\bullet$ radicals are generated by the direct photolysis of chlorinated compounds (Eqs 7–8).

$$CHCl_3 + hv \rightarrow CHCl_2^\bullet + Cl^\bullet \quad (7)$$

$$C_2HCl_3 + hv \rightarrow C_2HCl_2^\bullet + Cl^\bullet \quad (8)$$

They proposed that $Cl^\bullet$ sensitize the oxidation of chloroform and trichloroethylene (Eqs 9–11).

$$CHCl_3 + Cl^\bullet \rightarrow CCl_3^\bullet + HCl \quad (9)$$

$$C_2HCl_3 + Cl^\bullet \rightarrow C_2Cl_3^\bullet + HCl \quad (10)$$

$$C_2HCl_3 + Cl^\bullet \rightarrow C_2HCl_4^\bullet \quad (11)$$

The difference between $Cl^\bullet$ sensitized chloroform and trichloroethylene degradation are in their UV absorption spectra. Chloroform absorbs radiation at 185 nm, whereby trichloroethylene absorbs at 254 nm. Furthermore, it was confirmed that oxygen and chloroform compete for 185 nm photons (Eqs 12–14).

$$O_2 + hv \rightarrow O(^1D) + O(^3P) \quad \text{at 185 nm} \quad (12)$$

$$O(^1D) + O_2 \rightarrow O(^3P) + O_2 \quad (13)$$

$$O(^3P) + 2O_2 \rightarrow O_3 + O_2 \tag{14}$$

The $Cl^\bullet$ radicals do not react with C-Cl bond, thus Feiyan et al. (2002) proposed the following mechanism for carbon tetrachloride degradation (Eqs 15–18).

$$CCl_4 + h\nu \rightarrow Cl^\bullet + CCl_3^\bullet \tag{15}$$

$$2CCl_3^\bullet \rightarrow 2Cl^\bullet + CCl_2^\bullet \tag{16}$$

$$CCl_3^\bullet + O_2 \rightarrow CCl_3O_2^\bullet \tag{17}$$

$$2CCl_3O_2^\bullet \rightarrow 2COCl_2 + O_2 + 2Cl^\bullet \tag{18}$$

Alberici & Jardim (1997) observed no conversion of carbon tetrachloride by TiO_2 and irradiation with a black light lamp (365 nm) or germicidal lamp (254 nm). They slightly enhanced the photocatalytic degradation from 0 to 10% by addition of methanol as electron donor. Alberici & Jardim (1997) also investigated the degradation of nitrogen-containing compounds (pyridine). Consistent with other researchers the mineralization is slower in comparison to compounds with heteroatoms of chlorine. Zuo et al. (2007) studied the photolytic and photocatalytic degradation of cyanogen chloride by a germicidal lamp (254 nm). Neither by photolysis nor by photocatalysis and high intensive UV irradiation (10 mW/cm^2) the cyanogen chloride could be degraded. The photolysis of mono-, di- and trichloroamine was investigated by Li & Blatchley (2009). All three chloramines are degradable and the quantum yields are dependent on the wavelength. They examined quantum yields are greater than 1.0, which indicates that chain reactions are involved. The highest quantum yields were determined for monochloroamine with 3.46 mol/Es at 222 nm, for dichloroamine with 2.26 mol/Es at 222 nm and trichloroamine with 9.5 mol/Es at 282 nm.

2.2 *UV and photocatalytic treatment of bioaerosols*

The DNA molecule has its maximum absorption at the wavelength of 253.7 nm. Thus, ultraviolet germicidal irradiation ($\lambda = 254$ nm) is widely used. The helix structure is broken down which avoids the duplication of DNA as well as leads to cell mutation or lethality. The disadvantages of ultraviolet germicidal irradiation is the formation of ozone which is hazardous to human (Chen et al. 2010). In addition, the irradiation susceptibility of bacterial and fungal aerosols is influenced by relative humidity and irradiation dosage (Li et al. 2003). The ultraviolet germicidal irradiation dosage can differ by the factor of 80 between fungal spores and bacterial cells. Li et al. (2003) observed a very low dosage of ultraviolet germicidal irradiation for *Legionella pneumophila* of 289 to 860 μW sec/cm^2 to produce a 5 log decrease in concentration. In comparison, the dosages (germicidal UV) for 99% inactivation of other airborne germs are 1017 to 2356 μW sec/cm^2 for *E. coli*, 15949 to 19345 μW sec/cm^2 for *B. subtilis*, 12917 to 17497 μW sec/cm^2 for yeast and 47984 to 89419 μW sec/cm^2 for *P. citrinum*. If the same airborne germs were irradiated by a fluorescent black light (spectral maximum of 365 nm), only *Legionella pneumophila* showed sensitivity to UV-A with a survival efficiency of 0.85. The cell damage by long wavelength UV (320–400 nm) is explained by Oguma et al. (2002) through the exciting of photosensitive molecules producing active species, such as $O_2^{\bullet-}$ and •OH.

The photocatalytic destruction of airborne germs proceeds similar to the photocatalytic decomposition of organic matter. The difference is that cell membranes and layers are composed of different types of organic matter. Furthermore, germs posse protection and recovery mechanisms against photocatalytic oxidation (Chen et al. 2010). Viruses are the most sensitive microorganisms toward TiO_2 photocatalysis followed by Gram(−) bacteria, then Gram(+) bacteria and then by bacterial endospores, which are the most resistant (Josset et al. 2010b). Several mechanisms of germ inactivation are proposed (Chen et al. 2010). At first, the intracellular coenzyme A (CoA) is oxidized, which inhibits cell respiration and leads to cell death. The second mechanism suggests the decomposition of the cell wall followed by decomposition of the cell membrane until the whole organism is completely decomposed. This was observed by electron microscopy (Chen et al. 2010).

Josset et al. (2010a) constructed a 3D solid foam structured photocatalytic reactor and investigated the treatment of *Legionella pneumophila* contaminated airflows under UV-A irradiation. The airflow was 5 m^3/h with a flow residence time of 1.5 s. 94% of bacteria did not survive the

Table 1. Types and configuration of photoreactors used for gas phase treatment.

Type	Support material	Irradiation	Photo-catalyst	Ref.
Annular plug flow	glass cylinder	lamps with output at 365 nm or 254 nm	TiO_2	(Alberici & Jardim 1997)
	Pyrex glass tube	output at 300–400 nm	TiO_2	(Shiraishi et al. 2003, Shiraishi & Ishimatsu 2009)
Packed/ fluidized bed	silica	400 μW cm^{-2} at λmax 365 nm	TiO_2 and WO_3-TiO_2	(Hamill et al. 2001)
Packed bed	pellets	λ < 400 nm	TiO_2	(Yamazaki et al. 2005)
Fluidized bed	CeO, MgO, SiO_2, Al_2O_3, Geldart-C, 20 to 200 μm	UV	P25 TiO_2 + 0.1–5% Pt, Pd, Ag or Au	(Muggli 2007)
Monolith	porous Vycor glass	output at 350–550 nm	TiO_2	(Hwang et al. 1998)
	aluminium or aluminium alloy	output at 180–400 nm	TiO_2 + WO_3	(Wei et al. 2007)
	aluminium or aluminium alloy	output at 180–400 nm	Au/Pt-TiO_2 + MnO	(Wei et al. 2005)

single-pass through the reactor. Josset et al. compared the results with an annular TiO_2 coated reactor under same conditions and found that only 21% of *Legionella pneumophila* bacteria did not survive (Josset et al. 2010a). Thus, photocatalytic oxidation is a promising technology, if the process parameters like air flow rate, photocatalyst support structure, UV source and humidity are in the right range.

3 TECHNOLOGICAL ISSUES

3.1 *Photoreactors*

3.1.1 *General aspects*

Different reactor designs were proposed to provide optimal conditions for photocatalytic reactions. The photocatalyst has to be immobilized onto a support, which can be easily assembled to the photoreactor. The ideal catalyst support has the following properties (Pozzo et al. 1997):

a) strong bonding of the photocatalyst to the surface without negatively affecting their reactivity,
b) high specific area,
c) the support is not decomposed under UV irradiation,
d) good adsorption capability for compounds to be degraded,
e) facilitates mass transfer processes, and
f) is chemical inert.

Furthermore, the immobilization and fabrication shall be economical and technically realizable. Commonly the annular reactor, packed or fluidized bed reactor and monolithic reactor were proposed. A brief summary of laboratory and commercial photoreactors is given in Table 1.

3.1.2 *Annular plug flow photoreactor*

Annular plug flow photoreactors consist of a cylindrical tube (mainly Pyrex glass) where the inner surface is coated by photocatalyst (Alberici & Jardim 1997, Shiraishi et al. 2003, Shiraishi & Ishimatsu 2009). UV lamps are mounted in the center of the tube leaving a small gap between lamp and immobilized photocatalyst. This assembly provides high illumination intensity and high linear velocity of air passing through the reactor. For higher volume flows or longer retention times, several tubes have to be interconnected. A scheme of a typical annular plug flow reactor can be seen in Figure 1. The photocatalyst is immobilized by dip-coating technique. An aqueous

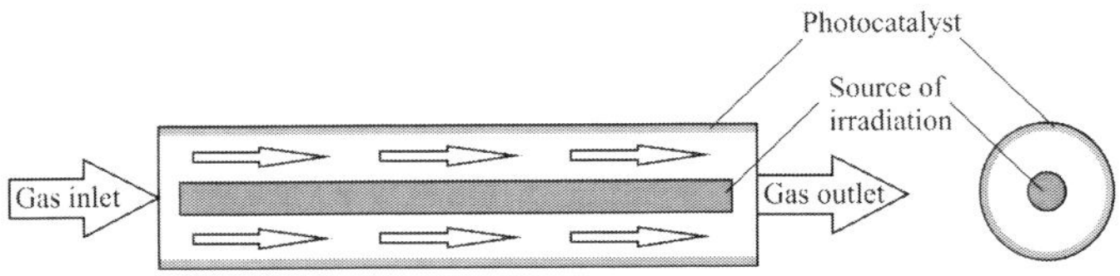

Figure 1. Typical annular plug flow photocatalytic reactor (Shiraishi et al. 2005).

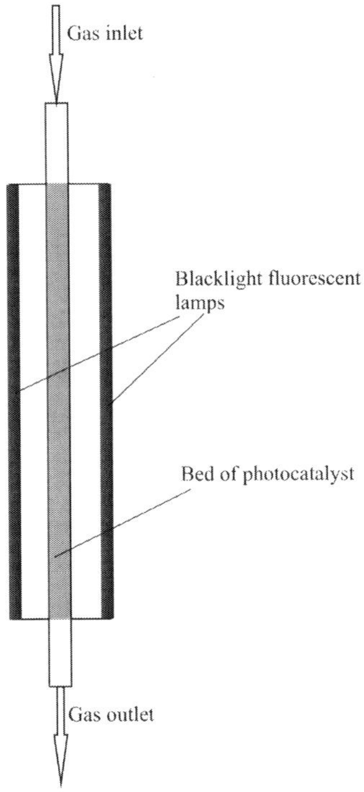

Figure 2. Packed bed photoreactor (Obuchi et al. 1999).

slurry or freshly prepared sol-gel is rinsed onto the internal tube surface, followed by drying and/or calcination.

3.1.3 *Packed bed photoreactor*

A typical assembly of packed bed photoreactor is shown in Figure 2. Photocatalyst pellets (Yamazaki et al. 2005) or photocatalyst immobilized on support material (Hamill et al. 2001) is filled into the reaction chamber. Hamill et al. prepared the photocatalyst by dip-coating of silica in photocatalyst suspension (Hamill et al. 2001). Also immersing of the support in sol-gel solution or chemical vapor deposition are further methods for immobilization of photocatalyst. The gas typically flows from the top to the bottom to prevent fluidization of the bed. The irradiation source can be mounted inside of the reactor or outside, if the reactor tube is transparent for UV light.

3.1.4 *Fluidized bed photoreactor*

In comparison to packed bed photoreactors, the gas stream in the fluidized bed photoreactor is passed from the bottom to the top. Also the support particles are smaller enabling fluidization. The mass transfer between gas phase and particle surface is improved. In addition, the photocatalyst is better exposed to irradiation enhancing the quantum yield. The disadvantage is attrition and discharge of photocatalyst. A typical assembly is shown in Figure 3. The lamps can be mounted inside or outside of the reactor (Hamill et al. 2001, Muggli 2007).

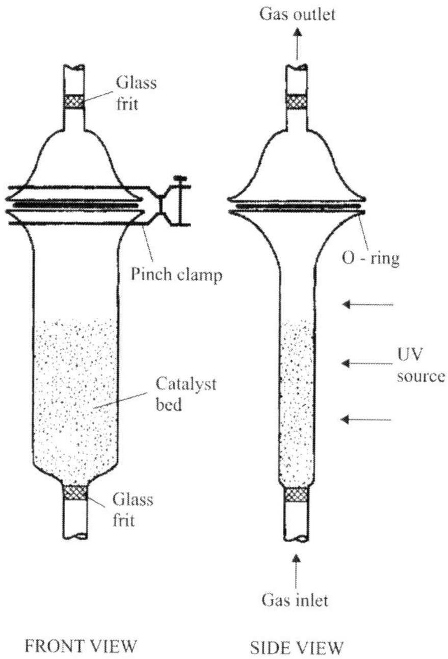

Figure 3. Fluidized bed photoreactor (Dibble & Raupp 1992).

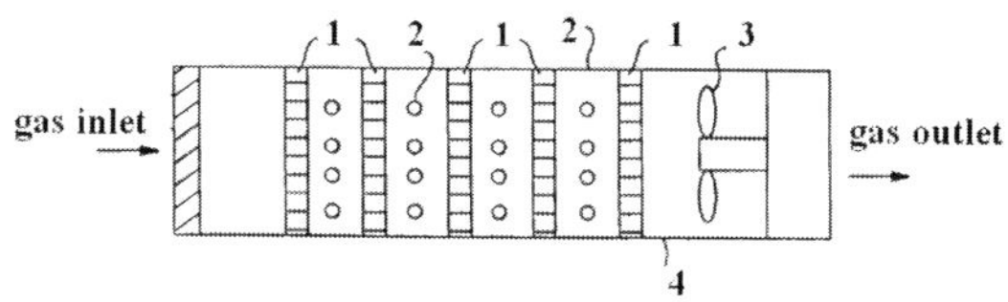

Figure 4. Monolith reactor with 1 monolith, 2 lamps, 3 fan and 4 reflecting housing material (Wei et al. 2007).

3.1.5 *Monolith reactor*

In technical application the monolith reactors are preferentially used (Vanderspurt et al. 2009, Wei et al. 2005, Wei et al. 2007). The monolith is made of aluminum, aluminum alloy or other suitable material in the shape of a honeycomb, square, rectangle or triangle. Monoliths offer well flow and irradiation conditions and are simple to produce. Honeycomb monolith photoreactors exhibit lower pressure drop than other types of photoreactors (Hossain 1998). In patents it is proposed to optimize the irradiation by lining the housing of the air purifier with reflecting material and to cascade monoliths and lamps (Wei et al. 2005, Wei et al. 2007). The photocatalysts are immobilized by dip coating, electrophoresis or spraying technique. Figure 4 shows a scheme of a monolith reactor.

3.2 *Effect of process parameters on photodegradation*

Photocatalytic degradation is influenced by various process parameters. Understanding the impact of these parameters is mandatory for scaling-up the process. The main parameters are gas velocity, ad- and desorption, irradiation, electron-hole-recombination, DBP, and oxygen concentration and humidity.

Photocatalytic reactions always involve adsorption steps. Several mechanisms where proposed such as Mars-Van Krevelen, stationary-state adsorption, Eley-Rideal, and Langmuir-Hinshelwood mechanism (Peral et al. 1997, Serpone & Pelizzetti 1989). The most used reaction mechanism to describe the heterogeneous photocatalytic decomposition of chlorinated compounds is the Langmuir-Hinshelwood mechanism (Demeestere et al. 2007, Mo et al. 2009). The degradation rate for a single component is expressed as (Eq.19):

$$r = \frac{kKC}{1+KC} \tag{19}$$

where k is the reaction rate constant and K is the adsorption equilibrium constant.

Oxygen concentration plays an essential role for the photocatalytic oxidation of chlorinated compounds (Borisch et al. 2004, Demeestere et al. 2007). Under anoxic conditions no significant degradation was observed (Demeestere et al. 2007). With increasing oxygen concentration the elimination rate increases and converges to a constant value. Yamazaki-Nishida et al. (1996) found that with increasing oxygen concentration the formation of unwanted byproducts was reduced. Dependent on the radiation source, reactive species of oxygen may be formed due to the 185 nm photon absorption (Eqs 12–14). Thus, also photolytic and homogeneous processes may be involved in the photodegradation.

The influence of relative humidity depends on its concentration as well as on the concentration and properties of the chlorinated DBPs. Water vapor competes with the DBPs for active sites and can decrease the elimination rate at increasing concentration (Demeestere et al. 2007, Mo et al. 2009). Several authors also report that water vapor suppresses the formation of phosgene (Hegedüs and Dombi 2004a, b, Hwang et al. 1998). Overall, water vapor retards the deactivation of the catalyst since less reactive surface-bound intermediates are formed (Demeestere et al. 2007). Paz (2010) finds in his extensive patent review that the sensitivity of the photocatalytic device to variations in humidity can be reduced by coating the titanium dioxide with a monolayer of tungsten oxide.

The gas velocity is the key parameter for the mass transfer of reactants to the surface. If the gas velocity is to slow, the degradation is limited by the transfer. With increasing velocity, the degradation rate will increase and converge. The velocity is also indirect proportional to the retention time. If the velocity is too high, the degradation rate will decrease since the reactant will leave the reactor before it could get adsorbed. It is similar for airborne germs. Josset et al. (2010b) investigated the influence of aerodynamics and found that viruses have lower probability to hit the photocatalyst surface due to their small size.

4 FINAL COMMENTS

This article highlights the photocatalytic air treatment in water-based recreational facilities and reviews how volatile chlorinated compounds and airborne microorganisms are decomposed. Up till now, papers on photocatalytical air treatment did not consider swimming pool conditions and requirements. In most investigations chloramines and halomethanes have been considered. Asthma cases in young children and swimmers correlate well with pool attendance. Microbiological threats including *Legionella* are of another concern. The largest exposure comes from inhaling contaminated air. Intense ventilation may not be sufficient, particularly when high ratio of air recirculation is considered. In this regard, advanced treatment of pool indoor air receives increasingly more attention. Degradation of contaminants is particularly important when most of the air is recirculated. Air recirculation means significant savings in energy consumption. Thus, photocatalytic air treatment is a promising technology.

ACKNOWLEDGEMENTS

This work was supported by the Voivodeship Fund for Environmental Protection and Water Management in Gdańsk, grant No. RX-10/2009, and the Gdańsk University of Technology, DS No. 014668/03.

REFERENCES

Alberici, R.M. & Jardim, W.F. 1997. Photocatalytic destruction of VOCs in the gas-phase using titanium dioxide. *Applied Catalysis B: Environmental* 14(1–2): 55–68.

Angenent, L.T., Kelley, S.T., Amand, A.S., Pace, N.R. & Hernandez, M.T. 2005. Molecular identification of potential pathogens in water and air of a hospital therapy pool. *Proceedings of the National Academy of Sciences of the United States of America* 102(13): 4860.

Borisch, J., Pilkenton, S., Miller, M.L., Raftery, D. & Francisco, J.S. 2004. TiO_2 photocatalytic degradation of dichloromethane: An FTIR and solid-state NMR study. *J. Phys. Chem. B* 108(18): 5640–5646.

Chen, F., Yang, X., Mak, H.K.C. & Chan, D.W.T. 2010. Photocatalytic oxidation for antimicrobial control in built environment: A brief literature overview. *Building and Environment* 45(8): 1747–1754.

Demeestere, K., Dewulf, J. & Van Langenhove, H. 2007. Heterogeneous photocatalysis as an advanced oxidation process for the abatement of chlorinated, monocyclic aromatic and sulfurous volatile organic compounds in air: State of the art. *Critical Reviews in Environmental Science and Technology* 37(6): 489–538.

Dibble, L.A. & Raupp, G.B. 1992. Fluidized-bed photocatalytic oxidation of trichloroethylene in contaminated air streams. *Environmental Science & Technology* 26(3): 492–495.

DIN Norm. 1997. Treatment of swimming pool water – "Aufbereitung von Schwimm- und Badebeckenwasser". DIN19643.

Fantuzzi, G., Righi, E., Predieri, G., Giacobazzi, P., Mastroianni, K. & Aggazzotti, G. 2010. Prevalence of Ocular, Respiratory and Cutaneous Symptoms in Indoor Swimming Pool Workers and Exposure to Disinfection By-Products (DBPs). *Int. J. Environ. Res. Public Health* 7(4):1379–91.

Feiyan, C., Pehkonen, S.O. & Ray, M.B. 2002. Kinetics and mechanisms of UV-photodegradation of chlorinated organics in the gas phase. *Water Research* 36(17): 4203–4214.

Hamill, N.A., Hardacre, C., Barth, J.A.C., Kalin, R.M. & Cunningham, J.F. 2001. Gas-phase photocatalytic oxidation of dichlorobutenes. *Environ. Sci. Technol.* 35(13): 2823–2827.

Hegedüs, M. & Dombi, A. 2004a. Comparative study of heterogeneous photocatalytic decomposition of tetrachloroethene and trichloroethene in the gas phase. *Applied Catalysis A: General* 271(1–2): 177–184.

Hegedüs, M. & Dombi, A. 2004b. Gas-phase heterogeneous photocatalytic oxidation of chlorinated ethenes over titanium dioxide: perchloroethene. *Applied Catalysis B: Environmental* 53(3): 141–151.

Hery, M., Hecht, G., Gerber, J.M., Gender, J.C., Hubert, G. & Rebuffaud, J. 1995. Exposure to chloramines in the atmosphere of indoor swimming pools. *Annals of Occupational Hygiene* 39(4): 427.

Hossain, M. 1998. Radiation field modeling in a photocatalytic monolith reactor. *Chemical Engineering Science* 53(22): 3771–3780.

Hwang, S.J., Petucci, C. & Raftery, D. 1998. In situ solid-state NMR studies of trichloroethylene photocatalysis: formation and characterization of surface-bound intermediates. *J. Am. Chem. Soc.* 120(18): 4388–4397.

Jacobs, J.H., Spaan, S., Van Rooy, G., Meliefste, C., Zaat, V.A.C., Rooyackers, J.M. & Heederik, D. 2007. Exposure to trichloramine and respiratory symptoms in indoor swimming pool workers. *European Respiratory Journal* 29(4): 690.

Josset, S., Hajiesmaili, S., Begin, D., Edouard, D., Pham-Huu, C., Lett, M.C., Keller, N. & Keller, V. 2010a. UV-A photocatalytic treatment of *Legionella pneumophila* bacteria contaminated airflows through three-dimensional solid foam structured photocatalytic reactors. *Journal of Hazardous Materials* 175(1–3): 372–381.

Josset, S., Taranto, J., Keller, N., Keller, V. & Lett, M.C. 2010b. Photocatalytic Treatment of Bioaerosols: Impact of the Reactor Design. *Environmental Science & Technology* 44(7): 2605–2611.

Leoni, E., Legnani, P.P., Bucci Sabattini, M.A. & Righi, F. 2001. Prevalence of *Legionella spp.* in swimming pool environment. *Water Research* 35(15): 3749–3753.

Li, C.S., Tseng, C.C., Lai, H.H. & Chang, C.W. 2003. Ultraviolet germicidal irradiation and titanium dioxide photocatalyst for controlling *Legionella pneumophila*. *Aerosol Science and Technology* 37(12): 961–966.

Li, J. & Blatchley III, E.R. 2009. UV Photodegradation of Inorganic Chloramines. *Environmental Science & Technology* 43(1): 60–65.

Mo, J., Zhang, Y., Xu, Q., Lamson, J.J. & Zhao, R. 2009. Photocatalytic purification of volatile organic compounds in indoor air: A literature review. *Atmospheric Environment* 43(14): 2229–2246.

Muggli, D.S. 2007. *Photocatalytic fluidized bed air purifier*. WO2007076134.

Nemery, B., Hoet, P.H.M. & Nowak, D. 2002. Indoor swimming pools, water chlorination and respiratory health. *European Respiratory Journal* 19(5): 790–793.

Obuchi, E., Sakamoto, T., Nakano, K. & Shiraishi, F. 1999. Photocatalytic decomposition of acetaldehyde over TiO_2/SiO_2 catalyst. *Chemical Engineering Science* 54(10): 1525–1530.

Oguma, K., Katayama, H. & Ohgaki, S. 2002. Photoreactivation of *Escherichia coli* after low-or medium-pressure UV disinfection determined by an endonuclease sensitive site assay. *Applied and Environmental Microbiology* 68(12): 6029–6035.

Paz, Y. 2010. Application of TiO_2 photocatalysis for air treatment: Patents' overview. *Applied Catalysis B: Environmental* 99: 448–460.
Peral, J., Domenech, X. & Ollis, D.F. 1997. Heterogeneous photocatalysis for purification, decontamination and deodorization of air. *Journal of Chemical Technology and Biotechnology* 70: 117–140.
Pozzo, R.L., Baltanas, M.A. & Cassano, A.E. 1997. Supported titanium oxide as photocatalyst in water decontamination: state of the art. *Catalysis Today* 39(3): 219–231.
Serpone, N. & Pelizzetti, E. 1989. *Photocatalysis: Fundamentals and Applications*, Wiley, New York, pp. 217–250,
Shiraishi, F., Yamaguchi, S. & Ohbuchi, Y. 2003. A rapid treatment of formaldehyde in a highly tight room using a photocatalytic reactor combined with a continuous adsorption and desorption apparatus. *Chemical Engineering Science* 58(3–6): 929–934.
Shiraishi, F., Ohkubo, D., Toyoda, K. & Yamaguchi, S. 2005. Decomposition of gaseous formaldehyde in a photocatalytic reactor with a parallel array of light sources: 1. Fundamental experiment for reactor design. *Chemical Engineering Journal* 114(1–3): 153–159.
Shiraishi, F. & Ishimatsu, T. 2009. Toluene removal from indoor air using a miniaturized photocatalytic air purifier including a preceding adsorption/desorption unit. *Chemical Engineering Science* 64(10): 2466–2472.
Vanderspurt, T.H., Davies, J.A., Hay, S.O., Obee, T.N., Opalka, S.M. & Wei, D. 2009. *Air purification system*. US20090246091.
Weaver, W.A., Li, J., Wen, Y., Johnston, J., Blatchley, M.R. & Blatchley Iii, E.R. 2009. Volatile disinfection by-product analysis from chlorinated indoor swimming pools. *Water Research* 43(13): 3308–3318.
Wei, D., Obee, T.N., Hay, S.O., Vanderspurt, T.H. & Schmidt, W.R. 2005. *Bifunctional layered photocatalyst/thermocatalyst for improving indoor air quality*. US20050129591.
Wei, D., Obee, T.N., Hay, S.O., Vanderspurt, T.H., Schmidt, W.R. & Sangiovanni, J.J. 2007. *Tungsten oxide/titanium dioxide photocatalyst for improving indoor air quality*. US7255831.
WHO. 2006. Guidelines for Safe Recreational-Water Environments, Vol. 2: Swimming Pools, Spas and Similar Recreational-Water Environments, World Health Organization, Geneva, Switzerland.
Yamazaki-Nishida, S., Fu, X., Anderson, M.A. & Hori, K. 1996. Chlorinated by products from the photoassisted catalytic oxidation of trichloroethylene and tetrachloroethylene in the gas phase using porous TiO_2 pellets. *Journal of Photochemistry and Photobiology A: Chemistry* 97(3): 175–179.
Yamazaki, S., Yoshida, A. & Abe, H. 2005. Photocatalytic degradation of chloroform in the gas phase on the porous TiO_2 pellets: effect of Cl accumulated on the catalyst surface. *Journal of Photochemistry and Photobiology A: Chemistry* 169(2): 191–196.
Zuo, G.M., Cheng, Z.X., Li, G.W., Shi, W.P. & Miao, T. 2007. Study on photolytic and photocatalytic decontamination of air polluted by chemical warfare agents (CWAs). *Chemical Engineering Journal* 128(2–3): 135–140.

Management of Indoor Air Quality – Dudzińska (ed)
© 2011 Taylor & Francis Group, London, ISBN 978-0-415-67266-5

Author index